Yamaha TZR125 & DT125 R Owners Workshop Manual

by Mark Coombs

Models covered
TZR125. 124cc. March 1987 to August 1993
DT125 R. 124cc. April 1988 on

Note: The TZR125 R, introduced in April 1993, is not covered by this Manual.

(1655 – 11T4)

Haynes Publishing
Sparkford Nr Yeovil
Somerset BA22 7JJ England

Haynes Publications, Inc
861 Lawrence Drive
Newbury Park
California 91320 USA

Acknowledgements

Our thanks are due to APS Motorcycles of Wells, Somerset, and Poole Motorcycles of Poole, Dorset, who supplied the machines featured in this Manual. Thanks are also due to Mitsui Machinery Sales (UK) Ltd who provided the necessary service information and gave permission to reproduce many of the line drawings used.

The Avon Rubber Company supplied information on tyre care and fitting, and NGK Spark Plugs (UK) Ltd provided information on plug maintenance and electrode conditions.

© Haynes Publishing 1995

A book in the **Haynes Owners Workshop Manual Series**

Printed by J. H. Haynes & Co. Ltd., Sparkford, Nr Yeovil, Somerset BA22 7JJ, England

All rights reserved. No part of this book may be reproduced or transmitted in any form or by any means, electronic or mechanical, including photocopying, recording or by any information storage or retrieval system, without permission in writing from the copyright holder.

ISBN 1 85960 030 1

British Library Cataloguing in Publication Data
A catalogue record for this book is available from the British Library

We take great pride in the accuracy of information given in this manual, but motorcycle manufacturers make alterations and design changes during the production run of a particular motorcycle of which they do not inform us. No liability can be accepted by the authors or publishers for loss, damage or injury caused by any errors in, or omissions from, the information given.

Restoring and Preserving our Motoring Heritage

Few people can have had the luck to realise their dreams to quite the same extent and in such a remarkable fashion as John Haynes, Founder and Chairman of the Haynes Publishing Group.

Since 1965 his unique approach to workshop manual publishing has proved so successful that millions of Haynes Manuals are now sold every year throughout the world, covering literally thousands of different makes and models of cars, vans and motorcycles.

A continuing passion for cars and motoring led to the founding in 1985 of a Charitable Trust dedicated to the restoration and preservation of our motoring heritage. To inaugurate the new Museum, John Haynes donated virtually his entire private collection of 52 cars.

Now with an unrivalled international collection of over 210 veteran, vintage and classic cars and motorcycles, the Haynes Motor Museum in Somerset is well on the way to becoming one of the most interesting Motor Museums in the world.

A 70 seat video cinema, a cafe and an extensive motoring bookshop, together with a specially constructed one kilometre motor circuit, make a visit to the Haynes Motor Museum a truly unforgettable experience.

Every vehicle in the museum is preserved in as near as possible mint condition and each car is run every six months on the motor circuit.

Enjoy the picnic area set amongst the rolling Somerset hills. Peer through the William Morris workshop windows at cars being restored, and browse through the extensive displays of fascinating motoring memorabilia.

From the 1903 Oldsmobile through such classics as an MG Midget to the mighty 'E' type Jaguar, Lamborghini, Ferrari Berlinetta Boxer, and Graham Hill's Lola Cosworth, there is something for everyone, young and old alike, at this Somerset Museum.

Haynes Motor Museum

Situated mid-way between London and Penzance, the Haynes Motor Museum is located just off the A303 at Sparkford, Somerset (home of the Haynes Manual) and is open to the public 7 days a week all year round, except Christmas Day and Boxing Day.

Telephone 01963 440804.

Contents

Introductory pages

About this manual	5
Introduction to the Yamaha TZR125 and DT125 R models	5
Model dimensions and weights	6
Ordering spare parts	6
Safety first!	7
Tools and working facilities	8
Spanner size comparison	10
Standard torque settings	10
Choosing and fitting accessories	11
Fault diagnosis	14
Routine maintenance	23

Chapter 1
Engine, clutch and gearbox   39

Chapter 2
Cooling system   71

Chapter 3
Fuel system and lubrication   78

Chapter 4
Ignition system   93

Chapter 5
Frame and forks   98

Chapter 6
Wheels, brakes and tyres   113

Chapter 7
Electrical system   131

Wiring diagrams   139

Conversion factors   144

Index   145

Yamaha TZR125 (3PC1) model

Yamaha DT125 R (3RN2) model

About this manual

The purpose of this manual is to present the owner with a concise and graphic guide which will enable him to tackle any operation from basic routine maintenance to a major overhaul. It has been assumed that any work would be undertaken without the luxury of a well-equipped workshop and a range of manufacturer's service tools.

To this end, the machine featured in the manual was stripped and rebuilt in our own workshop, by a team comprising a mechanic, a photographer and the author. The resulting photographic sequence depicts events as they took place, the hands shown being those of the author and the mechanic.

The use of specialised, and expensive, service tools was avoided unless their use was considered to be essential due to risk of breakage or injury. There is usually some way of improvising a method of removing a stubborn component, providing that a suitable degree of care is exercised.

The author learnt his motorcycle mechanics over a number of years, faced with the same difficulties and using similar facilities to those encountered by most owners. It is hoped that this practical experience can be passed on through the pages of this manual.

Where possible, a well-used example of the machine is chosen for the workshop project, as this highlights any areas which might be particularly prone to giving rise to problems. In this way, any such difficulties are encountered and resolved before the text is written, and the techniques used to deal with them can be incorporated in the relevant section. Armed with a working knowledge of the machine, the author undertakes a considerable amount of research in order that the maximum amount of data can be included in the manual.

A comprehensive section, preceding the main part of the manual, describes procedures for carrying out the routine maintenance of the machine at intervals of time and mileage. This section is included particularly for those owners who wish to ensure the efficient day-to-day running of their motorcycle, but who choose not to undertake overhaul or renovation work.

Each Chapter is divided into numbered sections. Within these sections are numbered paragraphs. Cross reference throughout the manual is quite straightforward and logical. When reference is made 'See Section 6.10' it means Section 6, paragraph 10 in the same Chapter. If another Chapter were intended, the reference would read, for example, 'See Chapter 2, Section 6.10'. All the photographs are captioned with a section/paragraph number to which they refer and are relevant to the Chapter text adjacent.

Figures (usually line illustrations) appear in a logical but numerical order, within a given Chapter. Fig. 1.1 therefore refers to the first figure in Chapter 1.

Left-hand and right-hand descriptions of the machines and their components refer to the left and right of a given machine when the rider is seated normally.

Motorcycle manufacturers continually make changes to specifications and recommendations, and these, when notified, are incorporated into our manuals at the earliest opportunity.

Introduction to the Yamaha TZR125 and DT125 R models

Before 1982 the 125cc capacity class of motorcycle was not particularly popular in the UK, since the machines were too slow to attract in large numbers the younger riders looking for performance, and yet were too large to be attractive to the commuter or non-enthusiast rider.

This situation altered radically in 1982 when new legislation dictated that all learner motorcyclists purchasing new motorcycles were to be restricted to machines of a maximum engine size of 125 cc, the power output being restricted to 9 kW (12.2 bhp). Almost immediately the main four Japanese manufacturers responded with a selection of new models to comply with this new legislation. Yamaha offered the RD125 LC and DT125 LC models. Both models used single cylinder, water cooled two-stroke engines and were equipped with Yamaha's Monocross rear suspension. They proved to be very successful and stayed in production until early 1987 and 1988 when they were superseded by the TZR125 and DT125 R models. The DT125 R model is very similar in appearance to its predecessor whereas the TZR125 model has been completely redesigned. The main design change to the TZR model was in the use of a 'Deltabox' type frame, constructed in thinwall steel. To complete the sports styling a full fairing is available from Yamaha as an optional extra.

Both the TZR125 and DT125 R employ a specially restricted version of the power valve (YPVS) engine sold in other markets. The power valve is fitted to the engine but is pegged in one position to comply with UK legislation. Since their introduction both models have been updated at regular intervals.

The original TZR125 model (code no 2RK) was superseded in 1989 by the 3PC1 model, although the only changes made were cosmetic with revised paintwork and graphics. It was not until the beginning of 1990 when the 3PC2 was introduced, that the TZR125 underwent any major change. The front brake caliper was modified and the rear drum brake of the earlier models was replaced with an hydraulically operated disc brake. Additionally, distinctive three-spoke alloy wheels were fitted, which could accept tubeless tyres. Apart from cosmetic changes, the 1991 TZR125 (model code 3PC3) was unchanged from its predecessor.

The original DT125 R (code 3DB1) model was replaced in 1989 by the 3RN1 model. Apart from new paintwork and graphics the only change was to the top end of the engine, together with a flat-slide carburettor. The 3RN1 model was superseded by the 3RN2 model at the beginning of 1990 and by the 3RN4 model at the beginning of 1991; differing only in colour and graphics. Later versions of the DT125 R (the 3RN6, 3RN7 and 3RN8) differed only in colour and graphics.

To help owners identify their machines exactly, the approximate dates of import are given below, with the initial engine and frame number with which each model's production run commenced. Note that where necessary, models are identified by their code numbers (eg 2RK) throughout this manual.

Model	Dates of import	Engine/frame no.
TZR125 2RK	Mar '87 to Mar '89	2RK-000101
TZR125 3PC1	Mar '89 to Feb '90	2RK-011101
TZR125 3PC2	Feb '90 to Dec '90	2RK-018101
TZR125 3PC3	Jan '91 to Aug '93	2RK-025101
DT125 R 3DB1	Apr '88 to May '89	3DB-000101
DT125 R 3RN1	May '89 to Feb '90	3MD-000101
DT125 R 3RN2	Feb '90 to Dec '90	3MD-008101
DT125 R 3RN4	Jan '91 to Feb '93	3MD-018101
DT125 R 3RN6	Feb '93 to Jan '94	3MD-036101
DT125 R 3RN7	Jan '94 to Nov '94	3MD-037101
DT125 R 3RN8	Nov '94 on	3MD-039101

Model dimensions and weights

	TZR model (2RK, 3PC1)	TZR model (3PC2, 3PC3)	DT model
Overall length	2020 mm (79.5 in)	2025 mm (79.7 in)	2160 mm (85.0 in)
Overall width	695 mm (27.4 in)	695 mm (27.4 in)	830 mm (32.7 in)
Overall height	1005 mm (39.6 in)	1005 mm (39.6 in)	1255 mm (49.4 in)
Seat height	760 mm (29.9 in)	765 mm (30.1 in)	885 mm (34.8 in)
Wheelbase	1340 mm (52.8 in)	1340 mm (52.8 in)	1415 mm (55.7 in)
Ground clearance	135 mm (5.3 in)	140 mm (5.5 in)	315 mm (12.4 in)
Kerb weight	120 kg (265 lb)	121 kg (267 lb)	119 kg (262 lb)

Ordering spare parts

When ordering spare parts it is advisable to deal direct with an authorized Yamaha dealer, who will be able to supply many of the items required ex-stock. It is advisable to get acquainted with the local Yamaha dealer, and to rely on his advice when purchasing spares. He is in a better position to specify exactly the parts required and to identify the relevant spare part numbers so that there is less chance of the wrong parts being supplied by the manufacturer due to a vague or incomplete description.

When ordering spares, always quote the frame and engine numbers in full, together with any prefixes or suffixes in the form of letters. The frame number is found stamped on the right-hand side of the steering head, in line with the forks. The engine number is stamped on the left-hand side of the crankcase, immediately behind the clutch lifting arm.

Use only parts of genuine Yamaha manufacture. A few pattern parts are available, sometimes at cheaper prices, but there is no guarantee they will give such good service as the originals they replace. Retain any worn or broken parts until the replacements have been obtained; they are sometimes needed as a pattern to help identify the correct replacement when design changes have been made during a production run.

Some of the more expendable parts such as spark plugs, bulbs, tyres, oils and greases etc., can be obtained from accessory shops and motor factors, who have convenient opening hours, and can often be found not far from home. It is also possible to obtain parts on a Mail Order basis from a number of specialists who advertise regularly in the motorcycle magazines.

Location of frame number

Location of engine number

Safety first!

Professional motor mechanics are trained in safe working procedures. However enthusiastic you may be about getting on with the job in hand, do take the time to ensure that your safety is not put at risk. A moment's lack of attention can result in an accident, as can failure to observe certain elementary precautions.

There will always be new ways of having accidents, and the following points do not pretend to be a comprehensive list of all dangers; they are intended rather to make you aware of the risks and to encourage a safety-conscious approach to all work you carry out on your vehicle.

Essential DOs and DON'Ts

DON'T start the engine without first ascertaining that the transmission is in neutral.
DON'T suddenly remove the filler cap from a hot cooling system – cover it with a cloth and release the pressure gradually first, or you may get scalded by escaping coolant.
DON'T attempt to drain oil until you are sure it has cooled sufficiently to avoid scalding you.
DON'T grasp any part of the engine, exhaust or silencer without first ascertaining that it is sufficiently cool to avoid burning you.
DON'T allow brake fluid or antifreeze to contact the machine's paintwork or plastic components.
DON'T syphon toxic liquids such as fuel, brake fluid or antifreeze by mouth, or allow them to remain on your skin.
DON'T inhale dust – it may be injurious to health (see *Asbestos* heading).
DON'T allow any spilt oil or grease to remain on the floor – wipe it up straight away, before someone slips on it.
DON'T use ill-fitting spanners or other tools which may slip and cause injury.
DON'T attempt to lift a heavy component which may be beyond your capability – get assistance.
DON'T rush to finish a job, or take unverified short cuts.
DON'T allow children or animals in or around an unattended vehicle.
DON'T inflate a tyre to a pressure above the recommended maximum. Apart from overstressing the carcase and wheel rim, in extreme cases the tyre may blow off forcibly.
DO ensure that the machine is supported securely at all times. This is especially important when the machine is blocked up to aid wheel or fork removal.
DO take care when attempting to slacken a stubborn nut or bolt. It is generally better to pull on a spanner, rather than push, so that if slippage occurs you fall away from the machine rather than on to it.
DO wear eye protection when using power tools such as drill, sander, bench grinder etc.
DO use a barrier cream on your hands prior to undertaking dirty jobs – it will protect your skin from infection as well as making the dirt easier to remove afterwards; but make sure your hands aren't left slippery. Note that long-term contact with used engine oil can be a health hazard.
DO keep loose clothing (cuffs, tie etc) and long hair well out of the way of moving mechanical parts.
DO remove rings, wristwatch etc, before working on the vehicle – especially the electrical system.
DO keep your work area tidy – it is only too easy to fall over articles left lying around.
DO exercise caution when compressing springs for removal or installation. Ensure that the tension is applied and released in a controlled manner, using suitable tools which preclude the possibility of the spring escaping violently.
DO ensure that any lifting tackle used has a safe working load rating adequate for the job.
DO get someone to check periodically that all is well, when working alone on the vehicle.
DO carry out work in a logical sequence and check that everything is correctly assembled and tightened afterwards.
DO remember that your vehicle's safety affects that of yourself and others. If in doubt on any point, get specialist advice.
IF, in spite of following these precautions, you are unfortunate enough to injure yourself, seek medical attention as soon as possible.

Asbestos

Certain friction, insulating, sealing, and other products – such as brake linings, clutch linings, gaskets, etc – contain asbestos. *Extreme care must be taken to avoid inhalation of dust from such products since it is hazardous to health.* If in doubt, assume that they *do* contain asbestos.

Fire

Remember at all times that petrol (gasoline) is highly flammable. Never smoke, or have any kind of naked flame around, when working on the vehicle. But the risk does not end there – a spark caused by an electrical short-circuit, by two metal surfaces contacting each other, by careless use of tools, or even by static electricity built up in your body under certain conditions, can ignite petrol vapour, which in a confined space is highly explosive.

Always disconnect the battery earth (ground) terminal before working on any part of the fuel or electrical system, and never risk spilling fuel on to a hot engine or exhaust.

It is recommended that a fire extinguisher of a type suitable for fuel and electrical fires is kept handy in the garage or workplace at all times. Never try to extinguish a fuel or electrical fire with water.

Note: *Any reference to a 'torch' appearing in this manual should always be taken to mean a hand-held battery-operated electric lamp or flashlight. It does **not** mean a welding/gas torch or blowlamp.*

Fumes

Certain fumes are highly toxic and can quickly cause unconsciousness and even death if inhaled to any extent. Petrol (gasoline) vapour comes into this category, as do the vapours from certain solvents such as trichloroethylene. Any draining or pouring of such volatile fluids should be done in a well ventilated area.

When using cleaning fluids and solvents, read the instructions carefully. Never use materials from unmarked containers – they may give off poisonous vapours.

Never run the engine of a motor vehicle in an enclosed space such as a garage. Exhaust fumes contain carbon monoxide which is extremely poisonous; if you need to run the engine, always do so in the open air or at least have the rear of the vehicle outside the workplace.

The battery

Never cause a spark, or allow a naked light, near the vehicle's battery. It will normally be giving off a certain amount of hydrogen gas, which is highly explosive.

Always disconnect the battery earth (ground) terminal before working on the fuel or electrical systems.

If possible, loosen the filler plugs or cover when charging the battery from an external source. Do not charge at an excessive rate or the battery may burst.

Take care when topping up and when carrying the battery. The acid electrolyte, even when diluted, is very corrosive and should not be allowed to contact the eyes or skin.

If you ever need to prepare electrolyte yourself, always add the acid slowly to the water, and never the other way round. Protect against splashes by wearing rubber gloves and goggles.

Mains electricity and electrical equipment

When using an electric power tool, inspection light etc, always ensure that the appliance is correctly connected to its plug and that, where necessary, it is properly earthed (grounded). Do not use such appliances in damp conditions and, again, beware of creating a spark or applying excessive heat in the vicinity of fuel or fuel vapour. Also ensure that the appliances meet the relevant national safety standards.

Ignition HT voltage

A severe electric shock can result from touching certain parts of the ignition system, such as the HT leads, when the engine is running or being cranked, particularly if components are damp or the insulation is defective. Where an electronic ignition system is fitted, the HT voltage is much higher and could prove fatal.

Tools and working facilities

The first priority when undertaking maintenance or repair work of any sort on a motorcycle is to have a clean, dry, well-lit working area. Work carried out in peace and quiet in the well-ordered atmosphere of a good workshop will give more satisfaction and much better results than can usually be achieved in poor working conditions. A good workshop must have a clean flat workbench or a solidly constructed table of convenient working height. The workbench or table should be equipped with a vice which has a jaw opening of at least 4 in (100 mm). A set of jaw covers should be made from soft metal such as aluminium alloy or copper, or from wood. These covers will minimise the marking or damaging of soft or delicate components which may be clamped in the vice. Some clean, dry, storage space will be required for tools, lubricants and dismantled components. It will be necessary during a major overhaul to lay out engine/gearbox components for examination and to keep them where they will remain undisturbed for as long as is necessary. To this end it is recommended that a supply of metal or plastic containers of suitable size is collected. A supply of clean, lint-free, rags for cleaning purposes and some newspapers, other rags, or paper towels for mopping up spillages should also be kept. If working on a hard concrete floor note that both the floor and one's knees can be protected from oil spillages and wear by cutting open a large cardboard box and spreading it flat on the floor under the machine or workbench. This also helps to provide some warmth in winter and to prevent the loss of nuts, washers, and other tiny components which have a tendency to disappear when dropped on anything other than a perfectly clean, flat, surface.

Unfortunately, such working conditions are not always available to the home mechanic. When working in poor conditions it is essential to take extra time and care to ensure that the components being worked on are kept scrupulously clean and to ensure that no components or tools are lost or damaged.

A selection of good tools is a fundamental requirement for anyone contemplating the maintenance and repair of a motor vehicle. For the owner who does not possess any, their purchase will prove a considerable expense, offsetting some of the savings made by doing-it-yourself. However, provided that the tools purchased meet the relevant national safety standards and are of good quality, they will last for many years and prove an extremely worthwhile investment.

To help the average owner to decide which tools are needed to carry out the various tasks detailed in this manual, we have compiled three lists of tools under the following headings: *Maintenance and minor repair, Repair and overhaul,* and *Specialized.* The newcomer to practical mechanics should start off with the simpler jobs around the vehicle. Then, as his confidence and experience grow, he can undertake more difficult tasks, buying extra tools as and when they are needed. In this way, a *Maintenance and minor repair* tool kit can be built-up into a *Repair and overhaul* tool kit over a considerable period of time without any major cash outlays. The experienced home mechanic will have a tool kit good enough for most repair and overhaul procedures and will add tools from the specialized category when he feels the expense is justified by the amount of use these tools will be put to.

It is obviously not possible to cover the subject of tools fully here. For those who wish to learn more about tools and their use there is a book entitled *Motorcycle Workshop Practice Manual* (Book no 1454) available from the publishers of this manual. It also provides an introduction to basic workshop practice which will be of interest to a home mechanic working on any type of motor vehicle.

As a general rule, it is better to buy the more expensive, good quality tools. Given reasonable use, such tools will last for a very long time, whereas the cheaper, poor quality, items will wear out faster and need to be renewed more often, thus nullifying the original saving. There is also the risk of a poor quality tool breaking while in use, causing personal injury or expensive damage to the component being worked on.

For practically all tools, a tool factor is the best source since he will have a very comprehensive range compared with the average garage or accessory shop. Having said that, accessory shops often offer excellent quality tools at discount prices, so it pays to shop around. There are plenty of tools around at reasonable prices, but always aim to purchase items which meet the relevant national safety standards. If in doubt, seek the advice of the shop proprietor or manager before making a purchase.

The basis of any toolkit is a set of spanners. While open-ended spanners with their slim jaws, are useful for working on awkwardly-positioned nuts, ring spanners have advantages in that they grip the nut far more positively. There is less risk of the spanner slipping off the nut and damaging it, for this reason alone ring spanners are to be preferred. Ideally, the home mechanic should acquire a set of each, but if expense rules this out a set of combination spanners (open-ended at one end and with a ring of the same size at the other) will provide a good compromise. Another item which is so useful it should be considered an essential requirement for any home mechanic is a set of socket spanners. These are available in a variety of drive sizes. It is recommended that the $\frac{1}{2}$-inch drive type is purchased to begin with as although bulkier and more expensive than the $\frac{3}{8}$-inch type, the larger size is far more common and will accept a greater variety of torque wrenches, extension pieces and socket sizes. The socket set should comprise sockets of sizes between 8 and 24 mm, a reversible ratchet drive, an extension bar of about 10 inches in length, a spark plug socket with a rubber insert, and a universal joint. Other attachments can be added to the set at a later date.

Maintenance and minor repair tool kit

Set of spanners 8 – 24 mm
Set of sockets and attachments
14 mm spark plug spanner with rubber insert
Adjustable spanner
C-spanner/pin spanner
Torque wrench (same size drive as sockets)
Set of screwdrivers (flat blade)
Set of screwdrivers (cross-head)
Set of Allen keys 4 – 10 mm
Impact screwdriver and bits
Ball pein hammer – 2 lb
Hacksaw (junior)
Self-locking pliers – Mole grips or vice grips
Pliers – combination
Pliers – needle nose

Tools and working facilities

Wire brush (small)
Soft-bristled brush
Tyre pump
Tyre pressure gauge
Tyre tread depth gauge
Oil can
Fine emery cloth
Funnel (medium size)
Drip tray
Grease gun
Set of feeler gauges
Brake bleeding kit
Strobe timing light
Continuity tester (dry battery and bulb)
Soldering iron and solder
Wire stripper or craft knife
PVC insulating tape
Assortment of split pins, nuts, bolts, and washers

Repair and overhaul toolkit

The tools in this list are virtually essential for anyone undertaking major repairs to a motorcycle and are additional to the tools listed above.

Plastic or rubber soft-faced mallet
Pliers – electrician's side cutters
Circlip pliers – internal (straight or right-angled tips are available)
Circlip pliers – external
Cold chisel
Centre punch
Pin punch
Scriber
Scraper (made from soft metal such as aluminium or copper)
Soft metal drift
Steel rule/straightedge
Assortment of files
Electric drill and bits
Wire brush (large)
Soft wire brush (similar to those used for cleaning suede shoes)
Sheet of plate glass
Hacksaw (large)
Valve grinding tool
Valve grinding compound (coarse and fine)
Stud extractor set (E-Z out)

Specialized tools

This is not a list of the tools made by the machine's manufacturer to carry out a specific task on a limited range of models. Occasional references are made to such tools in the text of this manual and, in general, an alternative method of carrying out the task without the manufacturer's tool is given where possible. The tools mentioned in this list are those which are not used regularly and are expensive to buy in view of their infrequent use. Where this is the case it may be possible to hire or borrow the tools against a deposit from a local dealer or tool hire shop. An alternative is for a group of friends or a motorcycle club to join in the purchase.

Flywheel rotor puller
Clutch holding tool
Piston ring compressor
Universal bearing puller
Cylinder bore honing attachment (for electric drill)
Micrometer set
Vernier calipers
Dial gauge set
Cylinder compression gauge
Vacuum gauge set
Multimeter
Dwell meter/tachometer

Care and maintenance of tools

Whatever the quality of the tools purchased, they will last much longer if cared for. This means in practice ensuring that a tool is used for its intended purpose; for example screwdrivers should not be used as a substitute for a centre punch, or as chisels. Always remove dirt or grease and any metal particles but remember that a light film of oil will prevent rusting if the tools are infrequently used. The common tools can be kept together in a large box or tray but the more delicate, and more expensive, items should be stored separately where they cannot be damaged. When a tool is damaged or worn out, be sure to renew it immediately. It is false economy to continue to use a worn spanner or screwdriver which may slip and cause expensive damage to the component being worked on.

Fastening systems

Fasteners, basically, are nuts, bolts and screws used to hold two or more parts together. There are a few things to keep in mind when working with fasteners. Almost all of them use a locking device of some type; either a lock washer, locknut, locking tab or thread adhesive. All threaded fasteners should be clean, straight, have undamaged threads and undamaged corners on the hexagon head where the spanner fits. Develop the habit of replacing all damaged nuts and bolts with new ones.

Rusted nuts and bolts should be treated with a rust penetrating fluid to ease removal and prevent breakage. After applying the rust penetrant, let it 'work' for a few minutes before trying to loosen the nut or bolt. Badly rusted fasteners may have to be chiseled off or removed with a special nut breaker, available at tool shops.

Flat washers and lock washers, when removed from an assembly should always be replaced exactly as removed. Replace any damaged washers with new ones. Always use a flat washer between a lock washer and any soft metal surface (such as aluminium), thin sheet metal or plastic. Special locknuts can only be used once or twice before they lose their locking ability and must be renewed.

If a bolt or stud breaks off in an assembly, it can be drilled out and removed with a special tool called an E-Z out. Most dealer service departments and motorcycle repair shops can perform this task, as well as others (such as the repair of threaded holes that have been stripped out).

Spanner size comparison

Jaw gap (in)	Spanner size
0.250	1/4 in AF
0.276	7 mm
0.313	5/16 in AF
0.315	8 mm
0.344	11/32 in AF; 1/8 in Whitworth
0.354	9 mm
0.375	3/8 in AF
0.394	10 mm
0.433	11 mm
0.438	7/16 in AF
0.445	3/16 in Whitworth; 1/4 in BSF
0.472	12 mm
0.500	1/2 in AF
0.512	13 mm
0.525	1/4 in Whitworth; 5/16 in BSF
0.551	14 mm
0.563	9/16 in BSF
0.591	15 mm
0.600	5/16 in Whitworth; 3/8 in BSF
0.625	5/8 in AF
0.630	16 mm
0.669	17 mm
0.686	11/16 in AF
0.709	18 mm
0.710	3/8 in Whitworth; 7/16 in BSF
0.748	19 mm
0.750	3/4 in AF
0.813	13/16 in AF
0.820	7/16 in Whitworth; 1/2 in BSF
0.866	22 mm
0.875	7/8 in AF
0.920	1/2 in Whitworth; 9/16 in BSF
0.938	15/16 in AF
0.945	24 mm
1.000	1 in AF
1.010	9/16 in Whitworth; 5/8 in BSF
1.024	26 mm
1.063	1 1/16 in AF; 27 mm
1.100	5/8 in Whitworth; 11/16 in BSF
1.125	1 1/8 in AF
1.181	30 mm
1.200	11/16 in Whitworth; 3/4 in BSF
1.250	1 1/4 in AF
1.260	32 mm
1.300	3/4 in Whitworth; 7/8 in BSF
1.313	1 5/16 in AF
1.390	13/16 in Whitworth; 15/16 in BSF
1.417	36 mm
1.438	1 7/16 in AF
1.480	7/8 in Whitworth; 1 in BSF
1.500	1 1/2 in AF
1.575	40 mm; 15/16 in Whitworth
1.614	41 mm
1.625	1 5/8 in AF
1.670	1 in Whitworth; 1 1/8 in BSF
1.688	1 11/16 in AF
1.811	46 mm
1.813	1 13/16 in AF
1.860	1 1/8 in Whitworth; 1 1/4 in BSF
1.875	1 7/8 in AF
1.969	50 mm
2.000	2 in AF
2.050	1 1/4 in Whitworth; 1 3/8 in BSF
2.165	55 mm
2.362	60 mm

Standard torque settings

Specific torque settings will be found at the end of the specifications section of each chapter. Where no figure is given, It should be secured according to the table below.

Fastener type (thread diameter)	kgf m	lbf ft
5mm bolt or nut	0.45 – 0.6	3.5 – 4.5
6 mm bolt or nut	0.8 – 1.2	6 – 9
8 mm bolt or nut	1.8 – 2.5	13 – 18
10 mm bolt or nut	3.0 – 4.0	22 – 29
12 mm bolt or nut	5.0 – 6.0	36 – 43
5 mm screw	0.35 – 0.5	2.5 – 3.6
6 mm screw	0.7 – 1.1	5 – 8
6 mm flange bolt	1.0 – 1.4	7 – 10
8 mm flange bolt	2.4 – 3.0	17 – 22
10 mm flange bolt	3.5 – 4.5	25 – 33

Choosing and fitting accessories

The range of accessories available to the modern motorcyclist is almost as varied and bewildering as the range of motorcycles. This Section is intended to help the owner in choosing the correct equipment for his needs and to avoid some of the mistakes made by many riders when adding accessories to their machines. It will be evident that the Section can only cover the subject in the most general terms and so it is recommended that the owner, having decided that he wants to fit, for example, a luggage rack or carrier, seeks the advice of several local dealers and the owners of similar machines. This will give a good idea of what makes of carrier are easily available, and at what price. Talking to other owners will give some insight into the drawbacks or good points of any one make. A walk round the motorcycles in car parks or outside a dealer will often reveal the same sort of information.

The first priority when choosing accessories is to assess exactly what one needs. It is, for example, pointless to buy a large heavy-duty carrier which is designed to take the weight of fully laden panniers and topbox when all you need is a place to strap on a set of waterproofs and a lunchbox when going to work. Many accessory manufacturers have ranges of equipment to cater for the individual needs of different riders and this point should be borne in mind when looking through a dealer's catalogues. Having decided exactly what is required and the use to which the accessories are going to be put, the owner will need a few hints on what to look for when making the final choice. To this end the Section is now sub-divided to cover the more popular accessories fitted. Note that it is in no way a customizing guide, but merely seeks to outline the practical considerations to be taken into account when adding aftermarket equipment to a motorcycle.

Fairings and windscreens

A fairing is possibly the single, most expensive, aftermarket item to be fitted to any motorcycle and, therefore, requires the most thought before purchase. Fairings can be divided into two main groups: front fork mounted handlebar fairings and windscreens, and frame mounted fairings.

The first group, the front fork mounted fairings, are becoming far more popular than was once the case, as they offer several advantages over the second group. Front fork mounted fairings generally are much easier and quicker to fit, involve less modification to the motorcycle, do not as a rule restrict the steering lock, permit a wider selection of handlebar styles to be used, and offer adequate protection for much less money than the frame mounted type. They are also lighter, can be swapped easily between different motorcycles, and are available in a much greater variety of styles. Their main disadvantages are that they do not offer as much weather protection as the frame mounted types, rarely offer any storage space, and, if poorly fitted or naturally incompatible, can have an adverse effect on the stability of the motorcycle.

The second group, the frame mounted fairings, are secured so rigidly to the main frame of the motorcycle that they can offer a substantial amount of protection to motorcycle and rider in the event of a crash. They offer almost complete protection from the weather and, if double-skinned in construction, can provide a great deal of useful storage space. The feeling of peace, quiet and complete relaxation encountered when riding behind a good full fairing has to be experienced to be believed. For this reason full fairings are considered essential by most touring motorcyclists and by many people who ride all year round. The main disadvantages of this type are that fitting can take a long time, often involving removal or modification of standard motorcycle components, they restrict the steering lock and they can add up to about 40 lb to the weight of the machine. They do not usually affect the stability of the machine to any great extent once the front tyre pressure and suspension have been adjusted to compensate for the extra weight, but can be affected by sidewinds.

The first thing to look for when purchasing a fairing is the quality of the fittings. A good fairing will have strong, substantial brackets constructed from heavy-gauge tubing; the brackets must be shaped to fit the frame or forks evenly so that the minimum of stress is imposed on the assembly when it is bolted down. The brackets should be properly painted or finished – a nylon coating being the favourite of the better manufacturers – the nuts and bolts provided should be of the same thread and size standard as is used on the motorcycle and be properly plated. Look also for shakeproof locking nuts or locking washers to ensure that everything remains securely tightened down. The fairing shell is generally made from one of two materials: fibreglass or ABS plastic. Both have their advantages and disadvantages, but the main consideration for the owner is that fibreglass is much easier to repair in the event of damage occurring to the fairing. Whichever material is used, check that it is properly finished inside as well as out, that the edges are protected by beading and that the fairing shell is insulated from vibration by the use of rubber grommets at all mounting points. Also be careful to check that the windscreen is retained by plastic bolts which will snap on impact so that the windscreen will break away and not cause personal injury in the event of an accident.

Having purchased your fairing or windscreen, read the manufacturer's fitting instructions very carefully and check that you have all the necessary brackets and fittings. Ensure that the mounting brackets are located correctly and bolted down securely. Note that some manufacturers use hose clamps to retain the mounting brackets; these should be discarded as they are convenient to use but not strong enough for the task. Stronger clamps should be substituted; car exhaust pipe clamps of suitable size would be a good alternative. Ensure that the front forks can turn through the full steering lock available without fouling the fairing. With many types of frame-mounted fairing the handlebars will have to be altered or a different type fitted and the steering lock will be restricted by stops provided with the fittings. Also check that the fairing does not foul the front wheel or mudguard, in any steering position, under full fork compression. Re-route any cables, brake pipes or electrical wiring which may snag on the fairing and take great care to protect all electrical connections, using insulating tape. If the manufacturer's instructions are followed carefully at every stage no serious problems should be encountered. Remember that hydraulic pipes that have been disconnected must be carefully re-tightened and the hydraulic system purged of air bubbles by bleeding.

Two things will become immediately apparent when taking a motorcycle on the road for the first time with a fairing – the first is the tendency to underestimate the road speed because of the lack of wind pressure on the body. This must be very carefully watched until one has grown accustomed to riding behind the fairing. The second thing is the alarming increase in engine noise which is an unfortunate but inevitable by-product of fitting any type of fairing or windscreen, and is caused by normal engine noise being reflected, and in some cases amplified, by the flat surface of the fairing.

Luggage racks or carriers

Carriers are possibly the commonest item to be fitted to modern motorcycles. They vary enormously in size, carrying capacity, and durability. When selecting a carrier, always look for one which is made specifically for your machine and which is bolted on with as few separate brackets as possible. The universal-type carrier, with its mass of brackets and adaptor pieces, will generally prove too weak to be of any real use. A good carrier should bolt to the main frame, generally using the two suspension unit top mountings and a mudguard mounting bolt as attachment points, and have its luggage platform as low and as far forward as possible to minimise the effect of any load on the machine's stability. Look for good quality, heavy gauge tubing, good welding and good finish. Also ensure that the carrier does not prevent opening of the seat, sidepanels or tail compartment, as appropriate. When using a carrier, be very careful not to overload it. Excessive weight placed so high and so far to the rear of any motorcycle will have an adverse effect on the machine's steering and stability.

Luggage

Motorcycle luggage can be grouped under two headings: soft and hard. Both types are available in many sizes and styles and have advantages and disadvantages in use.

Soft luggage is now becoming very popular because of its lower cost and its versatility. Whether in the form of tankbags, panniers, or strap-on bags, soft luggage requires in general no brackets and no modification to the motorcycle. Equipment can be swapped easily from one motorcycle to another and can be fitted and removed in seconds. Awkwardly shaped loads can easily be carried. The disadvantages of soft luggage are that the contents cannot be secure against the casual thief, very little protection is afforded in the event of a crash, and waterproofing is generally poor. Also, in the case of panniers, carrying capacity is restricted to approximately 10 lb, although this amount will vary considerably depending on the manufacturer's recommendation. When purchasing soft luggage, look for good quality material, generally vinyl or nylon, with strong, well-stitched attachment points. It is always useful to have separate pockets, especially on tank bags, for items which will be needed on the journey. When purchasing a tank bag, look for one which has a separate, well-padded, base. This will protect the tank's paintwork and permit easy access to the filler cap at petrol stations.

Hard luggage is confined to two types: panniers, and top boxes or tail trunks. Most hard luggage manufacturers produce matching sets of these items, the basis of which is generally that manufacturer's own heavy-duty luggage rack. Variations on this theme occur in the form of separate frames for the better quality panniers, fixed or quickly-detachable luggage, and in size and carrying capacity. Hard luggage offers a reasonable degree of security against theft and good protection against weather and accident damage. Carrying capacity is greater than that of soft luggage, around 15 – 20 lb in the case of panniers, although top boxes should never be loaded as much as their apparent capacity might imply. A top box should only be used for lightweight items, because one that is heavily laden can have a serious effect on the stability of the machine. When purchasing hard luggage look for the same good points as mentioned under fairings and windscreens, ie good quality mounting brackets and fittings, and well-finished fibreglass or ABS plastic cases. Again as with fairings, always purchase luggage made specifically for your motorcycle, using as few separate brackets as possible, to ensure that everything remains securely bolted in place. When fitting hard luggage, be careful to check that the rear suspension and brake operation will not be impaired in any way and remember that many pannier kits require re-siting of the indicators. Remember also that a non-standard exhaust system may make fitting extremely difficult.

Handlebars

The occupation of fitting alternative types of handlebar is extremely popular with modern motorcyclists, whose motives may vary from the purely practical, wishing to improve the comfort of their machines, to the purely aesthetic, where form is more important than function. Whatever the reason, there are several considerations to be borne in mind when changing the handlebars of your machine. If fitting lower bars, check carefully that the switches and cables do not foul the petrol tank on full lock and that the surplus length of cable, brake pipe, and electrical wiring are smoothly and tidily disposed of. Avoid tight kinks in cable or brake pipes which will produce stiff controls or the premature and disastrous failure of an overstressed component. If necessary, remove the petrol tank and re-route the cable from the engine/gearbox unit upwards, ensuring smooth gentle curves are produced. In extreme cases, it will be necessary to purchase a shorter brake pipe to overcome this problem. In the case of higher handlebars than standard it will almost certainly be necessary to purchase extended cables and brake pipes. Fortunately, many standard motorcycles have a custom version which will be equipped with higher handlebars and, therefore, factory-built extended components will be available from your local dealer. It is not usually necessary to extend electrical wiring, as switch clusters may be used on several different motorcycles, some being custom versions.

This point should be borne in mind however when fitting extremely high or wide handlebars.

When fitting different types of handlebar, ensure that the mounting clamps are correctly tightened to the manufacturer's specifications and that cables and wiring, as previously mentioned, have smooth easy runs and do not snag on any part of the motorcycle throughout the full steering lock. Ensure that the fluid level in the front brake master cylinder remains level to avoid any chance of air entering the hydraulic system. Also check that the cables are adjusted correctly and that all handlebar controls operate correctly and can be easily reached when riding.

Crashbars

Crashbars, also known as engine protector bars, engine guards, or case savers, are extremely useful items of equipment which can contribute protection to the machine's structure if a crash occurs. They do not, as has been inferred in the US, prevent the rider from crashing, or necessarily prevent rider injury should a crash occur.

It is recommended that only the smaller, neater, engine protector type of crashbar is considered. This type will offer protection while restricting, as little as is possible, access to the engine and the machine's ground clearance. The crashbars should be designed for use specifically on your machine, and should be constructed of heavy-gauge tubing with strong, integral mounting brackets. Where possible, they should bolt to a strong lug on the frame, usually at the engine mounting bolts.

The alternative type of crashbar is the larger cage type. This type is not recommended in spite of their appearance which promises some protection to the rider as well as to the machine. The larger amount of leverage imposed by the size of this type of crashbar increases the risk of severe frame damage in the event of an accident. This type also decreases the machine's ground clearance and restricts access to the engine. The amount of protection afforded the rider is open to some doubt as the design is based on the premise that the rider will stay in the normally seated position during an accident, and the crash bar structure will not itself fail. Neither result can in any way be guaranteed.

As a general rule, always purchase the best, ie usually the most expensive, set of crashbars you an afford. The investment will be repaid by minimising the amount of damage incurred, should the machine be involved in an accident. Finally, avoid the universal type of crashbar. This should be regarded only as a last resort to be used if no alternative exists. With its usual multitude of separate brackets and spacers, the universal crashbar is far too weak in design and construction to be of any practical value.

Exhaust systems

The fitting of aftermarket exhaust systems is another extremely popular pastime amongst motorcyclists. The usual motive is to gain more performance from the engine but other considerations are to gain more ground clearance, to lose weight from the motorcycle, to obtain a more distinctive exhaust note or to find a cheaper alternative to the manufacturer's original equipment exhaust system. Original equipment exhaust systems often cost more and may well have a relatively short life. It should be noted that it is rare for an aftermarket exhaust system alone to give a noticeable increase in the engine's power output. Modern motorcycles are designed to give the highest power output possible allowing for factors such as quietness, fuel economy, spread of power, and long-term reliability. If there were a magic formula which allowed the exhaust system to produce more power without affecting these other considerations you can be sure that the manufacturers, with their large research and development facilities, would have found it and made use of it. Performance increases of a worthwhile and noticeable nature only come from well-tried and properly matched modifications to the entire engine, from the air filter, through the carburettors, port timing or camshaft and valve design, combustion chamber shape, compression ratio, and the exhaust system. Such modifications are well outside the scope of this manual but interested owners might refer to specialist books produced by the publisher of this manual which go into the whole subject in great detail.

Whatever your motive for wishing to fit an alternative exhaust system, be sure to seek expert advice before doing so. Changes to the carburettor jetting will almost certainly be required for which you must consult the exhaust system manufacturer. If he cannot supply adequately specific information it is reasonable to assume that insufficient development work has been carried out, and that particular make should be avoided. Other factors to be borne in mind are whether the exhaust system allows the use of both centre and side stands, whether

Choosing and fitting accessories

it allows sufficient access to permit oil and filter changing and whether modifications are necessary to the standard exhaust system. Many two-stroke expansion chamber systems require the use of the standard exhaust pipe; this is all very well if the standard exhaust pipe and silencer are separate units but can cause problems if the two, swith so many modern two-strokes, are a one-piece unit. While the exhaust pipe can be removed easily by means of a hacksaw it is not so easy to refit the original silencer should you at any time wish to return the machine to standard trim. The same applies to several four-stroke systems.

On the subject of the finish of aftermarket exhausts, avoid black-painted systems unless you enjoy painting. As any trail-bike owner will tell you, rust has a great affinity for black exhausts and re-painting or rust removal becomes a task which must be carried out with monotonous regularity. A bright chrome finish is, as a general rule, a far better proposition as it is much easier to keep clean and to prevent rusting. Although the general finish of aftermarket exhaust systems is not always up to the standard of the original equipment the lower cost of such systems does at least reflect this fact.

When fitting an alternative system always purchase a full set of new exhaust gaskets, to prevent leaks. Fit the exhaust first to the cylinder head or barrel, as appropriate, tightening the retaining nuts or bolts by hand only and then line up the exhaust rear mountings. If the new system is a one-piece unit and the rear mountings do not line up exactly, spacers must be fabricated to take up the difference. Do not force the system into place as the stress thus imposed will rapidly cause cracks and splits to appear. Once all the mountings are loosely fixed, tighten the retaining nuts or bolts securely, being careful not to overtighten them. Where the motorcycle manufacturer's torque settings are available, these should be used. Do not forget to carry out any carburation changes recommended by the exhaust system's manufacturer.

Electrical equipment

The vast range of electrical equipment available to motorcyclists is so large and so diverse that only the most general outline can be given here. Electrical accessories vary from electronic ignition kits fitted to replace contact breaker points, to additional lighting at the front and rear, more powerful horns, various instruments and gauges, clocks, anti-theft systems, heated clothing, CB radios, radio-cassette players, and intercom systems, to name but a few of the more popular items of equipment.

As will be evident, it would require a separate manual to cover this subject alone and this section is therefore restricted to outlining a few basic rules which must be borne in mind when fitting electrical equipment. The first consideration is whether your machine's electrical system has enough reserve capacity to cope with the added demand of the accessories you wish to fit. The motorcycle's manufacturer or importer should be able to furnish this sort of information and may also be able to offer advice on uprating the electrical system. Failing this, a good dealer or the accessory manufacturer may be able to help. In some cases, more powerful generator components may be available, perhaps from another motorcycle in the manufacturer's range. The second consideration is the legal requirements in force in your area. The local police may be prepared to help with this point. In the UK for example, there are strict regulations governing the position and use of auxiliary riding lamps and fog lamps.

When fitting electrical equipment always disconnect the battery first to prevent the risk of a short-circuit, and be careful to ensure that all connections are properly made and that they are waterproof. Remember that many electrical accessories are designed primarily for use in cars and that they cannot easily withstand the exposure to vibration and to the weather. Delicate components must be rubber-mounted to insulate them from vibration, and sealed carefully to prevent the entry of rainwater and dirt. Be careful to follow exactly the accessory manufacturer's instructions in conjunction with the wiring diagram at the back of this manual.

Accessories – general

Accessories fitted to your motorcycle will rapidly deteriorate if not cared for. Regular washing and polishing will maintain the finish and will provide an opportunity to check that all mounting bolts and nuts are securely fastened. Any signs of chafing or wear should be watched for, and the cause cured as soon as possible before serious damage occurs.

As a general rule, do not expect the re-sale value of your motorcycle to increase by an amount proportional to the amount of money and effort put into fitting accessories. It is usually the case that an absolutely standard motorcycle will sell more easily at a better price than one that has been modified. If you are in the habit of exchanging your machine for another at frequent intervals, this factor should be borne in mind to avoid loss of money.

Fault diagnosis

Contents

Introduction	1

Engine does not start when turned over
No fuel flow to carburettor	2
Fuel not reaching cylinder	3
Engine flooding	4
No spark at plug	5
Weak spark at plug	6
Compression low	7

Engine stalls after starting
General causes	8

Poor running at idle and low speed
Weak spark at plug or erratic firing	9
Fuel/air mixture incorrect	10
Compression low	11

Acceleration poor
General causes	12

Poor running or lack of power at high speeds
Weak spark at plug or erratic firing	13
Fuel/air mixture incorrect	14
Compression low	15

Knocking or pinking
General causes	16

Overheating
Firing incorrect	17
Fuel/air mixture incorrect	18
Lubrication inadequate	19
Miscellaneous causes	20

Clutch operating problems
Clutch slip	21
Clutch drag	22

Gear selection problems
Gear lever does not return	23
Gear selection difficult or impossible	24
Jumping out of gear	25
Overselection	26

Abnormal engine noise
Knocking or pinking	27
Piston slap or rattling from cylinder	28
Other noises	29

Abnormal transmission noise
Clutch noise	30
Transmission noise	31

Exhaust smokes excessively
White/blue smoke (caused by oil burning)	32
Black smoke (caused by over-rich mixture)	33

Poor handling or roadholding
Directional instability	34
Steering bias to left or right	35
Handlebar vibrates or oscillates	36
Poor front fork performance	37
Front fork judder when braking	38
Poor rear suspension performance	39

Abnormal frame and suspension noise
Front end noise	40
Rear suspension noise	41

Brake problems
Brakes are spongy or ineffective – disc brakes	42
Brakes drag – disc brakes	43
Brake lever or pedal pulsates in operation – disc brakes	44
Disc brake noise	45
Brakes are spongy or ineffective – drum brakes	46
Brake drag – drum brakes	47
Brake lever or pedal pulsates in operation – drum brakes	48
Drum brake noise	49
Brake induced fork judder	50

Electrical problems
Battery dead or weak	51
Battery overcharged	52
Total electrical failure	53
Circuit failure	54
Bulbs blowing repeatedly	55

1 Introduction

This Section provides an easy reference-guide to the more common faults that are likely to afflict your machine. Obviously, the opportunities are almost limitless for faults to occur as a result of obscure failures, and to try and cover all eventualities would require a book. Indeed, a number have been written on the subject.

Successful fault diagnosis is not a mysterious 'black art' but the application of a bit of knowledge combined with a systematic and logical approach to the problem. Approach any fault diagnosis by first accurately identifying the symptom and then checking through the list of possible causes, starting with the simplest or most obvious and progressing in stages to the most complex. Take nothing for granted, but above all apply liberal quantities of common sense.

The main symptom of a fault is given in the text as a major heading below which are listed, as Section headings, the various systems or areas which may contain the fault. Details of each possible cause for a fault and the remedial action to be taken are given, in brief, in the paragraphs below each Section heading. Further information should be sought in the relevant Chapter.

Fault diagnosis

Engine does not start when turned over

2 No fuel flow to carburettor

● Fuel tank empty or level too low. Check that the tap is turned to 'On' or 'Reserve' position as required. If in doubt, prise off the fuel feed pipe at the carburettor end and check that fuel runs from the pipe when the tap is turned on.
● Tank filler cap vent obstructed. This can prevent fuel from flowing into the carburettor float chamber because air cannot enter the fuel tank to replace it. The problem is more likely to appear when the machine is being ridden. Check by listening close to the filler cap and releasing it. A hissing noise indicates that a blockage is present. Remove the cap and clear the vent hole with wire or by using an air line from the inside of the cap.
● Fuel tap or filter blocked. Blockage may be due to accumulation of rust or paint flakes from the tank's inner surface or of foreign matter from contaminated fuel. Remove the tap and clean it and the filter. Look also for water droplets in the fuel.
● Fuel pipe blocked. Blockage of the fuel pipe is more likely to result from a kink in the pipe rather than the accumulation of debris.

3 Fuel not reaching cylinder

● Float chamber not filling. Caused by float needle or floats sticking in up position. This may occur after the machine has been left standing for an extended length of time allowing the fuel to evaporate. When this occurs a gummy residue is often left which hardens to a varnish-like substance. This condition may be worsened by corrosion and crystalline deposits produced prior to the total evaporation of contaminated fuel. Sticking of the float needle may also be caused by wear. In any case removal of the float chamber will be necessary for inspection and cleaning.
● Blockage in starting circuit, slow running circuit or jets. Blockage of these items may be attributable to debris from the fuel tank by-passing the filter system or to gumming up as described in paragraph 1. Water droplets in the fuel will also block jets and passages. The carburettor should be dismantled for cleaning.
● Fuel level too low. The fuel level in the float chamber is controlled by float height. The fuel level may increase with wear or damage but will never reduce, thus a low fuel level is an inherent rather than developing condition. Check the float height, renewing the float or needle if required.

4 Engine flooding

● Float valve needle worn or stuck open. A piece of rust or other debris can prevent correct seating of the needle against the valve seat thereby permitting an uncontrolled flow of fuel. Similarly, a worn needle or needle seat will prevent valve closure. Dismantle the carburettor float chamber for cleaning and, if necessary, renewal of the worn components.
● Fuel level too high. The fuel level is controlled by the float height which may increase due to wear of the float needle, pivot pin or operating tang. Check the float height, and make any necessary adjustments. A leaking float will cause an increase in fuel level, and thus should be renewed.
● Cold starting mechanism. Check the choke (starter mechanism) for correct operation. If the mechanism jams in the 'On' position subsequent starting of a hot engine will be difficult.
● Blocked air filter. A badly restricted air filter will cause flooding. Check the filter and clean or renew as required. A collapsed inlet hose will have a similar effect. Check that the air filter inlet has not become blocked by a rag or similar item.

5 No spark at plug

● Ignition switch not on.
● Engine kill switch off.
● Fuse blown. See wiring diagram.
● Spark plug dirty, oiled or fouled. Because the induction mixture of a two-stroke engine is inclined to be of a rather oily nature it is comparatively easy to foul the plug electrodes, especially where there have been repeated attempts to start the engine. A machine used for short journeys will be more prone to fouling because the engine may never reach full operating temperature, and the deposits will not burn off. On rare occasions a change of plug grade may be required but the advice of a dealer should be sought before making such a change. On all two-stroke machines it is a sound precaution to carry a new spare spark plug for substitution in the event of fouling problems.
● Spark plug failure. Clean the spark plug thoroughly and reset the electrode gap. Refer to the spark plug section and the colour condition guide in Routine maintenance. If the spark plug shorts internally or has sustained visible damage to the electrodes, core or ceramic insulator it should be renewed. On rare occasions a plug that appears to spark vigorously will fail to do so when refitted to the engine and subjected to the compression pressure in the cylinder.
● Spark plug cap or high tension (HT) lead faulty. Check condition and security. Replace if deterioration is evident. Most spark plug caps have an internal resistor designed to inhibit electrical interference with radio and television sets. On rare occasions the resistor may break down, thus preventing sparking. If this is suspected, fit a new cap as a precaution.
● Spark plug cap loose. Check that the spark plug cap fits securely over the plug and, where fitted, the screwed terminal on the plug end is secure.
● Shorting due to moisture. Certain parts of the ignition system are susceptible to shorting when the machine is ridden or parked in wet weather. Check particularly the area from the spark plug cap back to the ignition coil. A water dispersant spray may be used to dry out waterlogged components. Recurrence of the problem can be prevented by using an ignition sealant spray after drying out and cleaning.
● Ignition or engine kill switch shorted. May be caused by water corrosion or wear. Water dispersant and contact cleaning sprays may be used. If this fails to overcome the problem dismantling and visual inspection of the switches will be required.
● Shorting or open circuit in wiring. Failure in any wire connecting any of the ignition components will cause ignition malfunction. Check also that all connections are clean, dry and tight.
● Ignition coil failure. Check the coil, referring to Chapter 4.
● CDI Unit faulty.

6 Weak spark at plug

● Feeble sparking at the plug may be caused by any of the faults mentioned in the preceding Section other than those items in the first three paragraphs. Check first the spark plug, this being the most likely culprit.

7 Compression low

● Spark plug loose. This will be self-evident on inspection, and may be accompanied by a hissing noise when the engine is turned over. Remove the plug and check that the threads in the cylinder head are not damaged. Check also that the plug sealing washer is in good condition.
● Cylinder head gasket leaking. This condition is often accompanied by a high pitched squeak from around the cylinder head and oil loss, and may be caused by insufficiently tightened cylinder head fasteners, a warped cylinder head or mechanical failure of the gasket material. Re-torquing the fasteners to the correct specification may seal the leak in some instances but if damage has occurred this course of action will provide, at best, only a temporary cure.
● Low crankcase compression. This can be caused by worn main bearings and seals and will upset the incoming fuel/air mixture. A good seal in these areas is essential on any two-stroke engine.

● Piston rings sticking or broken. Sticking of the piston rings may be caused by seizure due to lack of lubrication or overheating as a result of poor carburation or incorrect fuel type. Gumming of the rings may result from lack of use, or carbon deposits in the ring grooves. Broken rings result from over-revving, over-heating or general wear. In either case a top-end overhaul will be required.

Engine stalls after starting

8 General causes

● Improper cold start mechanism operation. Check that the operating controls function smoothly. A cold engine may not require application of an enriched mixture to start initially but may baulk without choke once firing. Likewise a hot engine may start with an enriched mixture but will stop almost immediately if the choke is inadvertently in operation.
● Ignition malfunction. See Section 9. Weak spark at plug.
● Carburettor incorrectly adjusted. Maladjustment of the mixture strength or idle speed may cause the engine to stop immediately after starting. See Chapter 3.
● Fuel contamination. Check for filter blockage by debris or water which reduces, but does not completely stop, fuel flow, or blockage of the slow speed circuit in the carburettor by the same agents. If water is present it can often be seen as droplets in the bottom of the float chamber. Clean the filter and, where water is in evidence, drain and flush the fuel tank and float chamber.
● Intake air leak. Check for security of the carburettor mounting and hose connections, and for cracks or splits in the hoses. Check also that the carburettor top is secure.
● Air filter blocked or omitted. A blocked filter will cause an over-rich mixture; the omission of a filter will cause an excessively weak mixture. Both conditions will have a detrimental effect on carburation. Clean or renew the filter as necessary.
● Fuel filler cap air vent blocked. Usually caused by dirt or water. Clean the vent orifice.
● Choked exhaust system. Caused by excessive carbon build-up in the system, particularly around the silencer baffles. Refer to Routine maintenance for further information.
● Excessive carbon build-up in the engine. This can result from failure to decarbonise the engine at the specified interval or through excessive oil consumption. Check pump adjustment.

Poor running at idle and low speed

9 Weak spark at plug or erratic firing

● Battery voltage low. In certain conditions low battery charge, especially when coupled with a badly sulphated battery, may result in misfiring. If the battery is in good general condition it should be recharged; an old battery suffering from sulphated plates should be renewed.
● Spark plug fouled, faulty or incorrectly adjusted. See Section 5 or refer to Routine maintenance.
● Spark plug cap or high tension lead shorting. Check the condition of both these items ensuring that they are in good condition and dry and that the cap is fitted correctly.
● Spark plug type incorrect. Fit plug of correct type and heat range as given in Specifications. In certain conditions a plug of hotter or colder type may be required for normal running.
● Ignition timing incorrect. See Chapter 4.
● Faulty ignition HT coil. Partial failure of the coil internal insulation will diminish the performance of the coil. No repair is possible, a new component must be fitted.
● Defective flywheel generator ignition source. Refer to Chapter 4 for further details on test procedures.

10 Fuel/air mixture incorrect

● Intake air leak. Check carburettor and air cleaner hoses for security and signs of splitting. Ensure that carburettor top is tight.
● Mixture strength incorrect. Adjust slow running mixture strength using pilot adjustment screw.
● Pilot jet or slow running circuit blocked. The carburettor should be removed and dismantled for thorough cleaning. Blow through all jets and air passages with compressed air to clear obstructions.
● Air cleaner clogged or omitted. Clean or fit air cleaner element as necessary. Check also that the element and air filter cover are correctly seated.
● Cold start mechanism in operation. Check that the choke has not been left on inadvertently and the operation is correct.
● Fuel level too high or too low. Check the float height, renewing float or needle if required. See Section 3 or 4.
● Fuel tank air vent obstructed. Obstructions usually caused by dirt or water. Clean vent orifice.

11 Compression low

● See Section 7.

Acceleration poor

12 General causes

● All items as for previous Section.
● Choked air filter. Failure to keep the air filter element clean will allow the build-up of dirt with proportional loss of performance. In extreme cases of neglect acceleration will suffer.
● Choked exhaust system. This can result from failure to remove accumulations of carbon from the silencer baffles at the prescribed intervals. The increased back pressure will make the machine noticeably sluggish. Refer to Routine maintenance for further information on decarbonisation.
● Excessive carbon build-up in the engine. This can result from failure to decarbonise the engine at the specified interval or through excessive oil consumption. Check pump adjustment.
● Ignition timing incorrect. As no provision for adjustment exists, test the electronic ignition components and renew as required.
● Carburation fault. See Section 10.
● Mechanical resistance. Check that the brakes are not binding. On small machines in particular note that the increased rolling resistance caused by under-inflated tyres may impede acceleration.

Poor running or lack of power at high speeds

13 Weak spark at plug or erratic firing

● All items as for Section 9.
● HT lead insulation failure. Insulation failure of the HT lead and spark plug cap due to old age or damage can cause shorting when the engine is driven hard. This condition may be less noticeable, or not noticeable at all at lower engine speeds.

14 Fuel/air mixture incorrect

● All items as for Section 10, with the exception of items relative exclusively to low speed running.
● Main jet blocked. Debris from contaminated fuel, or from the fuel tank, and water in the fuel can block the main jet. Clean the fuel filter, the float chamber area, and if water is present, flush and refill the fuel tank.
● Main jet is the wrong size. The standard carburettor jetting is for sea level atmospheric pressure. For high altitudes, usually above 5000 ft, a smaller main jet will be required.
● Jet needle and needle jet worn. These can be renewed individually but should be renewed as a pair. Renewal of both items requires partial dismantling of the carburettor.
● Air bleed holes blocked. Dismantle carburettor and use compressed air to blow out all air passages.

Fault diagnosis

- Reduced fuel flow. A reduction in the maximum fuel flow from the fuel tank to the carburettor will cause fuel starvation, proportionate to the engine speed. Check for blockages through debris or a kinked fuel pipe.

15 Compression low

- See Section 7.

Knocking or pinking

16 General causes

- Carbon build-up in combustion chamber. After high mileages have been covered a large accumulation of carbon may occur. This may glow red hot and cause premature ignition of the fuel/air mixture, in advance of normal firing by the spark plug. Cylinder head removal will be required to allow inspection and cleaning.
- Fuel incorrect. A low grade fuel, or one of poor quality may result in compression induced detonation of the fuel resulting in knocking and pinking noises. Old fuel can cause similar problems. A too highly leaded fuel will reduce detonation but will accelerate deposit formation in the combustion chamber and may lead to early pre-ignition as described in item 1.
- Spark plug heat range incorrect. Uncontrolled pre-ignition can result from the use of a spark plug the heat range of which is too hot.
- Weak mixture. Overheating of the engine due to a weak mixture can result in pre-ignition occurring where it would not occur when engine temperature was within normal limits. Maladjustment, blocked jets or passages and air leaks can cause this condition.

Overheating

17 Firing incorrect

- Spark plug fouled, defective or maladjusted. See Section 5.
- Spark plug type incorrect. Refer to the Specifications and ensure that the correct plug type is fitted.
- Incorrect ignition timing. See Chapter 4.

18 Fuel/air mixture incorrect

- Slow speed mixture strength incorrect. Adjust pilot air screw.
- Main jet wrong size. The carburettor is jetted for sea level atmospheric conditions. For high altitudes, usually above 5000 ft, a smaller main jet will be required.
- Air filter badly fitted or omitted. Check that the filter element is in place and that it and the air filter box cover are sealing correctly. Any leaks will cause a weak mixture.
- Induction air leaks. Check the security of the carburettor mountings and hose connections, and for cracks and splits in the hoses. Check also that the carburettor top is secure.
- Fuel level too low. See Section 3.
- Fuel tank filler cap air vent obstructed. Clear blockage.

19 Lubrication inadequate

- Oil pump settings incorrect. The oil pump settings are of great importance since the quantities of oil being injected are very small. Any variation in oil delivery will have a significant effect on the engine. Refer to Chapter 3 for further information.
- Oil tank empty or low. This will have disastrous consequences if left unnoticed. Check and replenish tank regularly.
- Transmission oil low or worn out. Check the level regularly and investigate any loss of oil. If the oil level drops with no sign of external leakage it is likely that the crankshaft main bearing oil seals are worn, allowing transmission oil to be drawn into the crankcase during induction.

20 Miscellaneous causes

- Radiator fins clogged. Accumulated debris in the radiator core will gradually reduce its ability to dissipate heat generated by the engine. It is worth noting that during the summer months dead insects can cause as many problems in this respect as road dirt and mud during the winter. Cleaning is best carried out by dislodging the debris with a high pressure hose from the back of the radiator. Once cleaned it is worth painting the matrix with a heat-dispersant matt black paint both to assist cooling and to prevent external corrosion. The fitting of some sort of mesh guard will help prevent the fins from becoming clogged, but make sure that this does not itself prevent adequate cooling.

Clutch operating problems

21 Clutch slip

- No clutch lever play. Adjust clutch lever end play according to the procedure in Routine maintenance.
- Friction plates worn or warped. Overhaul clutch assembly, replacing plates out of specification.
- Plain plates worn or warped. Overhaul clutch assembly, replacing plates out of specification.
- Clutch spring broken or worn. Old or heat-damaged (from slipping clutch) springs should be renewed as a set.
- Clutch release not adjusted properly. See the adjustments section of Routine maintenance.
- Clutch inner cable snagging. Caused by a frayed cable or kinked outer cable. Renew the cable. Repair of a frayed cable is not advised.
- Clutch release mechanism defective. Worn or damaged parts in the clutch release mechanism could include the lifting arm, return spring or pushrod. Renew parts as necessary.
- Clutch centre and outer drum worn. Severe indentation by the clutch plate tangs of the channels in the centre and drum will cause snagging of the plates preventing correct engagement. If this damage occurs, renewal of the worn components is required.
- Lubricant incorrect. Use of a transmission lubricant other than that specified may allow the plates to slip.

22 Clutch drag

- Clutch lever play excessive. Adjust lever at bars or at cable end if necessary.
- Clutch plates warped or damaged. This will cause a drag on the clutch, causing the machine to creep. Overhaul clutch assembly.
- Clutch spring tension uneven. Usually caused by a sagged or broken spring. Check and renew springs as a set.
- Transmission oil deteriorated. Badly contaminated transmission oil and a heavy deposit of oil sludge on the plates will cause plate sticking. The oil recommended for this machine is of the detergent type, therefore it is unlikely that this problem will arise unless regular oil changes are neglected.
- Transmission oil viscosity too high. Drag in the plates will result from the use of an oil with too high a viscosity. In very cold weather clutch drag may occur until the engine has reached operating temperature.
- Clutch centre and outer drum worn. Indentation by the clutch plate tangs of the channels in the centre and drum will prevent easy plate disengagement. If the damage is light the affected areas may be dressed with a fine file. More pronounced damage will necessitate renewal of the components.
- Clutch outer drum seized to shaft. Lack of lubrication, severe wear or damage can cause the drum to seize to the shaft. Overhaul of the clutch, and perhaps the transmission, may be necessary to repair damage.
- Clutch release mechanism defective. Worn or damaged release mechanism parts can stick and fail to provide leverage. Overhaul release components.

● Loose clutch nut. Causes drum and centre misalignment, putting a drag on the engine. Engagement adjustment continually varies. Overhaul clutch assembly.

Gear selection problems

23 Gear lever does not return

● Weak or broken return spring. Renew the spring.
● Gearchange shaft bent or seized. Distortion of the gearchange shaft often occurs if the machine is dropped heavily on the gear lever. Provided that damage is not severe straightening of the shaft is permissible.

24 Gear selection difficult or impossible

● Clutch not disengaging fully. See Section 22.
● Gearchange shaft bent. This often occurs if the machine is dropped heavily on the gear lever. Straightening of the shaft is permissible if the damage is not too great.
● Gearchange arm or pins worn or damaged. Wear or breakage of these items may cause difficulty in selecting one or more gears. Overhaul the selector mechanism.
● Selector drum detent arm damaged. Failure, rather than wear may jam the drum thereby preventing gearchanging or causing false selection at high speed.
● Selector forks bent or seized. This can be caused by dropping the machine heavily on the gearchange lever or as a result of lack of lubrication. Though rare, bending of a shaft can result from a missed gearchange or false selection at high speed.
● Selector fork end and pin wear. Pronounced wear of these items and the grooves in the gearchange drum can lead to imprecise selection and, eventually, no selection. Renewal of the worn components will be required.
● Structural failure. Failure of any one component of the selector rod and change mechanism will result in improper or fouled gear selection.

25 Jumping out of gear

● Detent arm assembly worn or damaged. Wear of the arm and the cam with which it locates and breakage of the return spring can cause imprecise gear selection resulting in jumping out of gear. Renew the damaged components.
● Gear pinion dogs worn or damaged. Rounding off the dog edges and the mating recesses in adjacent pinion can lead to jumping out of gear when under load. The gears should be inspected and renewed. Attempting to reprofile the dogs is not recommended.
● Selector forks, drum and pinion grooves worn. Extreme wear of these interconnected items can occur after high mileages especially when lubrication has been neglected. The worn components must be renewed.
● Gear pinions, bushes and shafts worn. Renew the worn components.
● Bent gearchange shaft. Often caused by dropping the machine on the gear lever.
● Gear pinion tooth broken. Chipped teeth are unlikely to cause jumping out of gear once the gear has been selected fully; a tooth which is completely broken off, however, may cause problems in this respect and in any event will cause transmission noise.

26 Overselection

● Detent arm worn or broken. Renew the damaged items.
● Gearchange arm stop pads worn. Repairs can be made by welding and reprofiling with a file.

Abnormal engine noise

27 Knocking or pinking

● See Section 16.

28 Piston slap or rattling from cylinder

● Cylinder bore/piston clearance excessive. Resulting from wear, or partial seizure. This condition can often be heard as a high, rapid tapping noise when the engine is under little or no load, particularly when power is just beginning to be applied. Reboring to the next correct oversize should be carried out and a new oversize piston fitted.
● Connecting rod bent. This can be caused by over-revving, trying to start a very badly flooded engine (resulting in an hydraulic lock in the cylinder) or by earlier mechanical failure. Attempts at straightening a bent connecting rod are not recommended. Careful inspection of the crankshaft should be made before renewing the damaged connecting rod.
● Gudgeon pin, piston boss bore or small-end bearing wear or seizure. Excess clearance or partial seizure between normal moving parts of these items can cause continuous or intermittent tapping noises. Rapid wear or seizure is caused by lubrication starvation.
● Piston rings worn, broken or sticking. Renew the rings after careful inspection of the piston and bore.

29 Other noises

● Big-end bearing wear. A pronounced knock from within the crankcase which worsens rapidly is indicative of big-end bearing failure as a result of extreme normal wear or lubrication failure. Remedial action in the form of a bottom end overhaul should be taken; continuing to run the engine will lead to further damage including the possibility of connecting rod breakage.
● Main bearing failure. Extreme normal wear or failure of the main bearings is characteristically accompanied by a rumble from the crankcase and vibration felt through the frame and footrests. Renew the worn bearings and carry out a very careful examination of the crankshaft.
● Crankshaft excessively out of true. A bent crank may result from over-revving or damage from an upper cylinder component or gearbox failure. Damage can also result from dropping the machine on either crankshaft end. Straightening of the crankshaft is not possible in normal circumstances; a replacement item should be fitted.
● Engine mounting loose. Tighten all the engine mounting nuts and bolts.
● Cylinder head gasket leaking. The noise most often associated with a leaking head gasket is a high pitched squeaking, although any other noise consistent with gas being forced out under pressure from a small orifice can also be emitted. Gasket leakage is often accompanied by oil seepage from around the mating joint or from the cylinder head holding down bolts and nuts. Leakage results from insufficient or uneven tightening of the cylinder head fasteners, or from random mechanical failure. Retightening to the correct torque figure will, at best, only provide a temporary cure. The gasket should be renewed at the earliest opportunity.
● Exhaust system leakage. Popping or crackling in the exhaust system, particularly when it occurs with the engine on the overrun, indicates a poor joint either at the cylinder port or at the exhaust pipe/silencer connection. Failure of the gasket or looseness of the clamp should be looked for.

Abnormal transmission noise

30 Clutch noise

● Clutch outer drum/friction plate tang clearance excessive.
● Clutch outer drum/thrust washer clearance excessive.
● Primary drive gear teeth worn or damaged.

Fault diagnosis

31 Transmission noise

- Bearing or bushes worn or damaged. Renew the affected components.
- Gear pinions worn or chipped. Renew the gear pinions.
- Metal chips jammed in gear teeth. This can occur when pieces of metal from any failed component are picked up by a meshing pinion. The condition will lead to rapid bearing wear or early gear failure.
- Transmission oil level too low. Top up immediately to prevent damage to gearbox and engine.
- Gearchange mechanism worn or damaged. Wear or failure of certain items in the selection and change components can induce mis-selection of gears (see Section 24) where incipient engagement of more than one gear set is promoted. Remedial action, by the overhaul of the gearbox, should be taken without delay.
- Chain snagging on cases or cycle parts. A badly worn chain or one that is excessively loose may snag or smack against adjacent components.

Exhaust smokes excessively

32 White/blue smoke (caused by oil burning)

- Oil pump settings incorrect. Check and reset the oil pump as described in Chapter 3.
- Crankshaft main bearing oil seals worn. Wear in the main bearing oil seals, often in conjunction with wear in the bearings themselves, can allow transmission oil to find its way into the crankcase and thence to the combustion chamber. This condition is often indicated by a mysterious drop in the transmission oil level with no sign of external leakage.
- Accumulated oil deposits in exhaust system. If the machine is used for short journeys only it is possible for the oil residue in the exhaust gases to condense in the relatively cool silencer. If the machine is then taken for a longer run in hot weather, the accumulated oil will burn off producing ominous smoke from the exhaust.

33 Black smoke (caused by over-rich mixture)

- Air filter element clogged. Clean or renew the element.
- Main jet loose or too large. Remove the float chamber to check for tightness of the jet. If the machine is used at high altitudes rejetting will be required to compensate for the lower atmospheric pressure.
- Cold start mechanism jammed on. Check that the mechanism works smoothly and correctly.
- Fuel level too high. The fuel level is controlled by the float height which can increase as a result of wear or damage. Remove the float chamber and check the float height. Check also that floats have not punctured; a punctured float will lose buoyancy and allow an increased fuel level.
- Float valve needle stuck open. Caused by dirt or a worn valve. Clean the float chamber or renew the needle and, if necessary, the valve seat.

Poor handling or roadholding

34 Directional instability

- Steering head bearing adjustment too tight. This will cause rolling or weaving at low speeds. Re-adjust the bearings.
- Steering head bearing worn or damaged. Correct adjustment of the bearing will prove impossible to achieve if wear or damage has occurred. Inconsistent handling will occur including rolling or weaving at low speed and poor directional control at indeterminate higher speeds. The steering head bearing should be dismantled for inspection and renewed if required. Lubrication should also be carried out.
- Bearing races pitted or dented. Impact damage caused, perhaps, by an accident or riding over a pot-hole can cause indentation of the bearing, usually in one position. This should be noted as notchiness when the handlebars are turned. Renew and lubricate the bearings.
- Steering stem bent. This will occur only if the machine is subjected to a high impact such as hitting a kerb or a pot-hole. The lower yoke/stem should be renewed; do not attempt to straighten the stem.
- Front or rear tyre pressures too low.
- Front or rear tyre worn. General instability, high speed wobbles and skipping over white lines indicates that tyre renewal may be required. Tyre induced problems, in some machine/tyre combinations, can occur even when the tyre in question is by no means fully worn.
- Swinging arm or linkage bearings/bushes worn. Difficulty in holding line, particularly when cornering or when changing power settings indicates wear in the bearings or bushes. The swinging arm/linkage should be removed from the machine and the bearings/bushes renewed.
- Swinging arm flexing. The symptoms given in the preceding paragraph will also occur if the swinging arm fork flexes badly. This can be caused by structural weakness as a result of corrosion, fatigue or impact damage, or because the rear wheel spindle is slack.
- Wheel bearings worn. Renew the worn bearings.
- Loose wheel spokes – DT model. The spokes should be tightened evenly to maintain tension and trueness of the rim.
- Tyres unsuitable for machine. Not all available tyres will suit the characteristics of the frame and suspension, indeed, some tyres or tyre combinations may cause a transformation in the handling characteristics. If handling problems occur immediately after changing to a new tyre type or make, revert to the original tyres to see whether an improvement can be noted. In some instances a change to what are, in fact, suitable tyres may give rise to handling deficiencies. In this case a thorough check should be made of all frame and suspension items which affect stability.

35 Steering bias to left or right

- Rear wheel out of alignment. Caused by uneven adjustment of chain tensioner adjusters allowing the wheel to be askew in the fork ends. A bent rear wheel spindle will also misalign the wheel in the swinging arm.
- Wheels out of alignment. This can be caused by impact damage to the frame, swinging arm, wheel spindles or front forks. Although occasionally a result of material failure or corrosion it is usually as a result of a crash.
- Front forks twisted in the steering yokes. A light impact, for instance with a pot-hole or low curb, can twist the fork legs in the steering yokes without causing structural damage to the fork legs or the yokes themselves. Re-alignment can be made by loosening the yoke pinch bolts, wheel spindle and mudguard bolts. Re-align the wheel with the handlebars and tighten the bolts working upwards from the wheel spindle. This action should be carried out only when there is no chance that structural damage has occurred.

36 Handlebar vibrates or oscillates

- Tyres worn or out of balance. Either condition, particularly in the front tyre, will promote shaking of the fork assembly and thus the handlebars. A sudden onset of shaking can result if a balance weight is displaced during use.
- Tyres badly positioned on the wheel rims. A moulded line on each wall of a tyre is provided to allow visual verification that the tyre is correctly positioned on the rim. A check can be made by rotating the tyre; any misalignment will be immediately obvious.
- Wheel rims warped or damaged. Inspect the wheels for runout as described in Chapter 6.
- Swinging arm or linkage bearings/bushes worn. Renew the bearings/bushes.
- Wheel bearings worn. Renew the bearings.
- Steering head bearings incorrectly adjusted. Vibration is more likely to result from bearings which are too loose rather than too tight. Re-adjust the bearings.
- Loose fork component fasteners. Loose nuts and bolts holding the fork legs, wheel spindle, mudguards or steering stem can promote shaking at the handlebars. Fasteners on running gear such as the forks and suspension should be check tightened occasionally to prevent

dangerous looseness of components occurring.
- Engine mounting bolts loose. Tighten all fasteners.

37 Poor front fork performance

- Damping fluid level incorrect. If the fluid level is too low poor suspension control will occur resulting in a general impairment of roadholding and early loss of tyre adhesion when cornering and braking. Too much oil is unlikely to change the fork characteristics unless severe overfilling occurs when the fork action will become stiffer and oil seal failure may occur.
- Damping oil viscosity incorrect. The damping action of the fork is directly related to the viscosity of the damping oil. The lighter the oil used, the less will be the damping action imparted. For general use, use the recommended viscosity of oil, changing to a slightly higher or heavier oil only when a change in damping characteristic is required. Overworked oil, or oil contaminated with water which has found its way past the seals, should be renewed to restore the correct damping performance and to prevent bottoming of the forks.
- Damping components worn or corroded. Advanced normal wear of the fork internals is unlikely to occur until a very high mileage has been covered. Continual use of the machine with damaged oil seals which allows the ingress of water, or neglect, will lead to rapid corrosion and wear. Dismantle the forks for inspection and overhaul.
- Weak fork springs. Progressive fatigue of the fork springs, resulting in a reduced spring free length, will occur after extensive use. This condition will promote excessive fork dive under braking, and in its advanced form will reduce the at-rest extended length of the forks and thus the fork geometry. Renewal of the springs as a pair is the only satisfactory course of action.
- Bent stanchions or corroded stanchions. Both conditions will prevent correct telescoping of the fork legs, and in an advanced state can cause sticking of the fork in one position. In a mild form corrosion will cause stiction of the fork thereby increasing the time the suspension takes to react to an uneven road surface. Bent fork stanchions should be attended to immediately because they indicate that impact damage has occurred, and there is a danger that the forks will fail with disastrous consequences.

38 Front fork judder when braking (see also Section 50)

- Wear between the fork stanchions and the fork legs. Renewal of the affected components is required.
- Slack steering head bearings. Re-adjust the bearings.
- Warped brake disc. If irregular braking action occurs fork judder can be induced in what are normally serviceable forks. Renew the damaged brake components.

39 Poor rear suspension performance

- Rear suspension unit damper worn out or leaking. The damping performance of most rear suspension units falls off with age. This is a gradual process, and thus may not be immediately obvious. Indications of poor damping include hopping of the rear end when cornering or braking, and a general loss of positive stability.
- Weak rear spring. If the suspension unit spring fatigues it will promote excessive pitching of the machine and reduce the ground clearance when cornering.
- Swinging arm flexing or bearings worn. See Sections 34 and 36.
- Suspension linkage pivot bearings or bushes worn. Overhaul as described in Chapter 5 – DT models.
- Bent suspension unit damper rod. This is likely to occur only if the machine is dropped or if seizure of the piston occurs. If either happens the suspension unit should be renewed.

Abnormal frame and suspension noise

40 Front end noise

- Oil level low or too thin. This can cause a 'spurting' sound and is usually accompanied by irregular fork action.
- Spring weak or broken. Makes a clicking or scraping sound. Fork oil will have a lot of metal particles in it.
- Steering head bearings loose or damaged. Clicks when braking. Check, adjust or renew.
- Fork clamps loose. Make sure all fork clamp pinch bolts are tight.
- Fork stanchion bent. Good possibility if machine has been dropped. Repair or replace stanchion.

41 Rear suspension noise

- Fluid level too low. Leakage of a suspension unit, usually evident by oil on the outer surfaces, can cause a spurting noise. The suspension unit should be renewed.
- Defective rear suspension unit with internal damage. Renew the suspension unit.

Brake problems

42 Brakes are spongy or ineffective – disc brakes

- Air in brake circuit. This is only likely to happen in service due to neglect in checking the fluid level or because a leak has developed. The problem should be identified and the brake system bled of air.
- Pad worn. Check the pad wear against the wear indicators provided and renew the pads if necessary.
- Contaminated pads. Cleaning pads which have been contaminated with oil, grease or brake fluid is unlikely to prove successful; the pads should be renewed.
- Pads glazed. This is usually caused by overheating. The surface of the pads may be roughened using glass-paper or a fine file.
- Brake fluid deterioration. A brake which on initial operation is firm but rapidly becomes spongy in use may be failing due to water contamination of the fluid. The fluid should be drained and then the system refilled and bled.
- Master cylinder seal failure. Wear or damage of master cylinder internal parts will prevent pressurisation of the brake fluid. Overhaul the master cylinder unit.
- Caliper seal failure. This will almost certainly be obvious by loss of fluid, a lowering of fluid in the master cylinder reservoir and contamination of the brake pads and caliper. Overhaul the caliper assembly.
- Brake lever or pedal improperly adjusted. Adjust the clearance as described in Routine maintenance.

43 Brake drag – disc brakes

- Disc warped. The disc must be renewed.
- Caliper piston, caliper or pads corroded. The brake caliper assembly is vulnerable to corrosion due to water and dirt, and unless cleaned at regular intervals and lubricated in the recommended manner, will become sticky in operation.
- Piston seal deteriorated. The seal is designed to return the piston in the caliper to the retracted position when the brake is released. Wear or old age can affect this function. The caliper should be overhauled if this occurs.
- Brake pad damaged. Pad material separating from the backing plate due to wear or faulty manufacture. Renew the pads. Faulty installation of a pad also will cause dragging.
- Wheel spindle bent. The spindle may be straightened if no structural damage has occurred.
- Brake lever or pedal not returning. Check that the lever or pedal works smoothly throughout its operating range and does not snag on any adjacent cycle parts. Lubricate the pivot if necessary.

Fault diagnosis

● Twisted caliper support bracket. This is likely to occur only after impact in an accident. No attempt should be made to re-align the caliper; the bracket should be renewed.

44 Brake lever or pedal pulsates in operation – disc brakes

● Disc warped or irregularly worn. The disc must be renewed.
● Wheel spindle bent. The spindle may be straightened provided no structural damage has occurred.

45 Disc brake noise

● Brake squeal. This can be caused by the omission or incorrect installation of the anti-squeal shims. Squealing can also be caused by dust on the pads, usually in combination with glazed pads, or other contamination from oil, grease, brake fluid or corrosion. Persistent squealing which cannot be traced to any of the normal causes can often be cured by applying a thin layer of high temperature silicone grease to the rear of the pads. Make absolutely certain that no grease is allowed to contaminate the braking surface of the pads.
● Glazed pads. This is usually caused by high temperatures or contamination. The pad surfaces may be roughened using glass-paper or a fine file. If this approach does not effect a cure the pads should be renewed.
● Disc warped. This can cause a chattering, clicking or intermittent squeal and is usually accompanied by a pulsating brake lever or pedal or uneven braking. The disc must be renewed.
● Brake pads fitted incorrectly or undersize. Longitudinal play in the pads due to omission of the anti-rattle springs (where fitted) or because pads of the wrong size have been fitted will cause a single tapping noise every time the brake is operated. Inspect the pads for correct installation and security.

46 Brakes are spongy or ineffective – drum brakes

● Worn brake linings. Determine lining wear using the external brake wear indicator on the brake backplate, or by removing the wheel and withdrawing the brake backplate. Renew the shoe/lining units as a pair if the linings are worn below the recommended limit.
● Worn brake camshaft. Wear between the camshaft and the bearing surface will reduce brake feel and reduce operating efficiency. Renewal of one or both items will be required to rectify the fault.
● Worn brake cam and shoe ends. Renew the worn components.
● Linings contaminated with dust or grease. Any accumulations of dust should be cleaned from the brake assembly and drum using a petrol dampened cloth. Do not blow or brush off the dust because it is asbestos based and thus harmful if inhaled. Light contamination from grease can be removed from the surface of the brake linings using a solvent; attempts at removing heavier contamination are less likely to be successful because some of the lubricant will have been absorbed by the lining material which will severely reduce the braking performance.

47 Brake drag – drum brakes

● Incorrect adjustment. Re-adjust the brake operating mechanism.
● Drum warped or oval. This can result from overheating or impact or uneven tension of the wheel spokes. The condition is difficult to correct, although if slight ovality only occurs, skimming the surface of the brake drum can provide a cure. This is work for a specialist engineer. Renewal of the complete wheel is normally the only satisfactory solution.
● Weak brake shoe return springs. This will prevent the brake lining/shoe units from pulling away from the drum surface once the brake is released. The springs should be renewed.
● Brake camshaft or pedal pivot poorly lubricated. Failure to attend to regular lubrication of these areas will increase operating resistance which, when compounded, may cause tardy operation and poor release movement.

48 Brake pedal pulsates in operation – drum brakes

● Drum warped or oval. This can result from overheating or impact or uneven spoke tension. This condition is difficult to correct, although if slight ovality only occurs skimming the surface of the drum can provide a cure. This is work for a specialist engineer. Renewal of the wheel is normally the only satisfactory solution.

49 Drum brake noise

● Drum warped or oval. This can cause intermittent rubbing of the brake linings against the drum. See the preceding Section.
● Brake linings glazed. This condition, usually accompanied by heavy lining dust contamination, often induces brake squeal. The surface of the linings may be roughened using glass-paper or a fine file.

50 Brake induced fork judder

● Worn front fork stanchions and legs, or worn or badly adjusted steering head bearings. These conditions, combined with uneven or pulsating braking as described in Sections 44 and 48 will induce more or less judder when the brakes are applied, dependent on the degree of wear and poor brake operation. Attention should be given to both areas of malfunction. See the relevant Sections.

Electrical problems

51 Battery dead or weak

● Battery faulty. Battery life should not be expected to exceed 3 to 4 years. Gradual sulphation of the plates and sediment deposits will reduce the battery performance. Plate and insulator damage can often occur as a result of vibration. Complete power failure, or intermittent failure, may be due to a broken battery terminal. Lack of electrolyte will prevent the battery maintaining charge.
● Battery leads making poor contact. Remove the battery leads and clean them and the terminals, removing all traces of corrosion and tarnish. Reconnect the leads and apply a coating of petroleum jelly to the terminals.
● Load excessive. If additional items such as spot lamps, are fitted, which increase the total electrical load above the maximum alternator output, the battery will fail to maintain full charge. Reduce the electrical load to suit the electrical capacity.
● Rectifier failure.
● Alternator generating coils open-circuit or shorted.
● Charging circuit shorting or open circuit. This may be caused by frayed or broken wiring, dirty connectors or a faulty ignition switch. The system should be tested in a logical manner. See Section 54.

52 Battery overcharged

● Regulator faulty. Overcharging is indicated if the battery becomes hot or it is noticed that the electrolyte level falls repeatedly between checks. In extreme cases the battery will boil causing corrosive gases and electrolyte to be emitted through the vent pipes.
● Battery wrongly matched to the electrical circuit. Ensure that the specified battery is fitted to the machine.

53 Total electrical failure

● Fuse blown. Check the main fuse. If a fault has occurred, it must be rectified before a new fuse is fitted.
● Battery faulty. See Section 51.
● Earth failure. Check that the main earth strap is securely affixed to the frame and is making a good contact.

● Ignition switch or power circuit failure. Check for current flow through the battery positive lead (red) to the ignition switch. Check the ignition switch for continuity.

54 Circuit failure

● Cable failure. Refer to the machine's wiring diagram and check the circuit for continuity. Open circuits are a result of loose or corroded connections, either at terminals or in-line connectors, or because of broken wires. Occasionally, the core of a wire will break without there being any apparent damage to the outer plastic cover.
● Switch failure. All switches may be checked for continuity in each switch position, after referring to the switch position boxes incorporated in the wiring diagram for the machine. Switch failure may be a result of mechanical breakage, corrosion or water.
● Fuse blown. Refer to the wiring diagram to check whether or not a circuit fuse is fitted. Replace the fuse, if blown, only after the fault has been identified and rectified.

55 Bulbs blowing repeatedly

● Vibration failure. This is often an inherent fault related to the natural vibration characteristics of the engine and frame and is, thus, difficult to resolve. Modifications of the lamp mounting, to change the damping characteristics, may help.
● Intermittent earth. Repeated failure of one bulb, particularly where the bulb is fed directly from the generator, indicates that a poor earth exists somewhere in the circuit. Check that a good contact is available at each earthing point in the circuit.
● Reduced voltage. Where a quartz-halogen bulb is fitted the voltage to the bulb should be maintained or early failure of the bulb will occur. Do not overload the system with additional electrical equipment in excess of the system's power capacity and ensure that all circuit connections are maintained clean and tight.

Routine maintenance

Specifications

Engine

	TZR models	DT models
Spark plug:		
NGK	BR8ES or BR9ES	BR9ES
Nippon Denso	W24ESR-U or W27ESR-U	W27ESR-U
Spark plug electrode gap	0.7 – 0.8 mm (0.028 – 0.031 in)	0.7 – 0.8 mm (0.028 – 0.031 in)
Idle speed	1300 – 1400 rpm	1300 – 1400 rpm
Throttle cable free play – at twistgrip flange	2 – 5 mm (0.08 – 0.20 in)	2 – 5 mm (0.08 – 0.20 in)
Clutch cable free play – at handlebar lever butt end	2 – 3 mm (0.08 – 0.12 in)	2 – 3 mm (0.08 – 0.12 in)

Cycle parts

	TZR models	DT models
Front brake lever free play – at lever tip	2 – 5 mm (0.08 – 0.20 in)	2 – 5 mm (0.08 – 0.20 in)
Rear brake pedal height – below top of footrest	41 mm (1.6 in)	15 mm (0.6 in)
Rear brake pedal freeplay – at pedal tip (drum brake models)	20 – 30 mm (0.8 – 1.2 in)	Not applicable
Final drive chain free play	30 – 35 mm (1.2 – 1.4 in)	25 – 40 mm (1.0 – 1.6 in)
Brake pad friction material service limit	0.5 mm (0.02 in)	0.8 mm (0.03 in)
Brake shoe friction material service limit (drum brake models)	2.0 mm (0.08 in)	Not applicable
Tyre pressures – tyres cold:		
Front (solo) up to 90 kg (198 lb) load	25 psi (1.75 kg/cm^2)	18 psi (1.26 kg/cm^2)
Rear (solo) up to 90 kg (198 lb) load	28 psi (1.96 kg/cm^2)	22 psi (1.54 kg/cm^2)
Front (pillion) 90 kg – max load	25 psi (1.75 kg/cm^2)	22 psi (1.54 kg/cm^2)
Rear (pillion) 90 kg – max load	32 psi (2.24 kg/cm^2)	26 psi (1.82 kg/cm^2)
Maximum load:		
TZR models	193 kg (425 lb)	
DT models		front 47 kg (104 lb), rear 134 kg (295 lb)

Note: For off-road riding, DT model pressures should be 18 psi (front) and 22 psi (rear)

Recommended lubricants and fluids

Fuel grade	Unleaded, minimum octane rating 91 (RON/RM)
Engine lubrication:	
Oil tank capacity	1.2 lit (2.1 Imp pt)
Recommended oil	Yamaha 2T oil or equivalent good quality air-cooled 2-stroke engine oil
Transmission lubrication:	
Capacity:	
At oil change	750 cc (1.3 Imp pt)
After rebuild (dry)	800 cc (1.4 Imp pt)
Recommended oil	SAE 10W/30 SE motor oil
Coolant	See Chapter 2
Front forks – capacity per leg:	
TZR models	238 cc (8.4 Imp fl oz)
Early DT model (3DB1)	486 cc (17.1 Imp fl oz)
Later DT models (3RN1 onward)	495 cc (17.5 Imp fl oz)
Recommended oil	10W fork oil
Fork oil level:	
TZR models	149 mm (5.87 in)
Early DT model (3DB1)	175 mm (6.89 in)
Later DT models (3RN1 onward)	165.5 mm (6.52 in)
Drive chain lubrication:	
Standard type chain	SAE 10W/30 SE motor oil or aerosol chain lubricant
O-ring type chain	SAE 30 – 50W motor oil or aerosol chain lubricant suitable for O-ring chains
Brake fluid	DOT 4 (if not available DOT 3 may be used)
Air filter	SAE 10W/30 SE motor oil
Steering head, wheel, swinging arm and rear suspension linkage pivot bearings	Good quality high melting-point lithium-based grease
Instrument drive cables and stand pivot	General purpose grease
Control cables and all other pivots	Engine oil or light machine oil

Torque settings

Component	TZR models kgf m	lbf ft	DT models kgf m	lbf ft
Spark plug	2.0	14	2.0	14
Brake caliper mounting bolts	3.5	25	Not applicable	
Brake pad retaining pins (2RK and 3PC1 models)	1.0	7.2	Not applicable	
Brake caliper body retaining bolt	Not applicable		1.8	13
Front wheel spindle	7.4	53	5.8	42
Rear wheel spindle nut	8.5	61	9.0	65
Transmission oil drain plug	1.5	11	1.5	11
Coolant drain plug	1.5	11	1.5	11

Introduction

Periodic routine maintenance is a continuous process which should commence immediately the machine is used. The object is to maintain all adjustments and to diagnose and rectify minor defects before they develop into more extensive, and often more expensive, problems.

It follows that if the machine is maintained properly, it will both run and perform with optimum efficiency, and be less prone to unexpected breakdowns. Regular inspection of the machine will show up any parts which are wearing and, and with a little experience, it is possible to obtain the maximum life from any one component, renewing it when it becomes so worn that it is liable to fail.

Regular cleaning can be considered as important as mechanical maintenance. This will ensure that all the cycle parts are inspected regularly and are kept free from accumulations of road dirt and grime.

All tasks are grouped under various mileage headings, all of which are also given calendar-based intervals; if the machine only covers a low mileage maintenance should be carried out according to the calendar headings instead. All intervals are intended as a guide only; as a machine gets older it develops individual faults which require more frequent attention and if used under particularly arduous conditions it is advisable to reduce the period between each check.

For ease of reference, most service operations are described in detail under the relevant heading. However, if further general information is required, this can be found under the pertinent Section heading and Chapter in the main text.

Although no special tools are required for routine maintenance, a good selection of general workshop tools is essential. Included in the tools must be a range of metric ring or combination spanners, a selection of crosshead screwdrivers, and two pairs of circlip pliers, one external opening and the other internal opening. One further item of equipment that can be regarded as essential for owners of these models is a stand of some sort that can hold the machine securely in an upright position with enough height to permit the wheels to be removed. This can be anything from an old milk crate to a purpose-built paddock-type metal stand.

Daily (pre-ride) checks

It is recommended that the following items are checked whenever the machine is about to be used. This is important to prevent the risk of unexpected failure of any component while riding the machine and, with experience, can be reduced to a simple checklist which will only take a few moments to complete. For those owners who are not inclined to check all items with such frequency, it is suggested that the best course is to carry out the checks in the form of a service which can be undertaken each week or before any long journey. It is essential that all items are checked and serviced with reasonable frequency.

1 Check the engine oil level

Although a warning lamp is fitted to give prior warning that the oil level is running low, this may not allow sufficient reserve to allow you to find some more. Accordingly it is recommended that the tank be kept topped up at all times; the true level can be seen through the translucent plastic of the tank.

On TZR models unlock and remove the seat to gain access to the filler cap. On DT models the oil tank is mounted on the right-hand side of the frame just behind the steering head. Unscrew the cap and add the specified grade of oil up to the bottom of the filler neck. Tighten the cap securely on refitting. It is useful to keep a spare container of oil with the machine so that a supply is always available.

Note: *If the low oil warning indicator illuminates when riding the machine ensure that the oil tank is replenished without delay.* In the event of the oil level falling low enough for air to enter the pipe between the oil tank and pump, the system must be bled before the machine is ridden. The bleeding operation is described in Chapter 3.

2 Check the coolant level

Although the cooling system is semi-sealed and should not require frequent topping up, it is still necessary to check the level at regular intervals. A separate expansion tank is fitted to allow for expansion of the coolant when the engine is hot, the displaced liquid being drawn back into the system when it cools. It is therefore the level of coolant in the expansion tank which is to be checked; the tank is constructed of translucent plastic so that the coolant level can be seen easily in relation to the upper and lower level lines marked on its side. *Check the coolant level only when the engine is cold.*

On TZR models the expansion tank is situated under the right-hand sidepanel, access to the tank can be gained by removing the seat and sidepanel. On DT models it will be necessary to remove the left-hand sidepanel to check the expansion tank. The coolant level must be between the higher ('Full' or 'F') and lower ('Low' or 'L') level marks at all times. If the level is significantly above the higher level mark at any time the surplus should be siphoned off to prevent coolant being blown over the rear of the machine via the tank breather.

Use only the specified ingredients to make coolant of the required strength, as described in Chapter 2, and always have a supply prepared for topping up. In cases of real emergency distilled water or **clean** rainwater may be used, but remember that this will dilute the coolant and reduce the degree of protection against freezing.

If the level falls steadily, check the system very carefully for leaks, as described in Chapter 2. Also, do not forget to check that the radiator matrix is clean, unblocked and free from damage of any sort; this applies particularly to owners of DT models who have been riding off-road.

3 Check the fuel level

Checking the petrol level may seem obvious, but it is all too easy to forget. Ensure that you have enough petrol to complete your journey, or at least to get you to the nearest petrol station.

4 Check the battery

The battery is located on the right-hand side of the machine, underneath the sidepanel. The seat (TZR only) and sidepanel must be removed to check the electrolyte level.

On all models, whenever the battery is disconnected, remember to disconnect the negative (–) terminal first, to prevent the possibility of short circuits. The electrolyte level, visible through the translucent casing, must be between the two level marks. If necessary remove the cell caps and top up to the upper level using only distilled water. Check that the terminals are clean and apply a thin smear of petroleum jelly (not grease) to each to prevent corrosion. On refitting, check that the vent hose is not blocked and that it is correctly routed with no kinks, also that it hangs well below any other component, particularly the chain or exhaust system. Secure the battery and fuse holder with the rubber strap. Remember always to connect the negative (–) terminal last when refitting the battery.

Always check that the terminals are tight and that the rubber covers are correctly refitted, also that the fuse connections are clean and tight,

Routine maintenance

Use only good quality two-stroke oil when topping up – TZR model shown

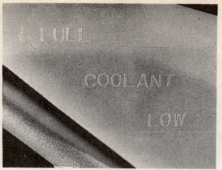
Coolant level should be between the marks on side of expansion tank – TZR model shown

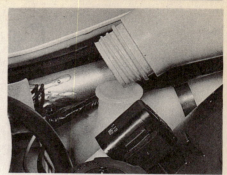
If level is low top up expansion tank

Electrolyte level must be between marks on battery casing

Check fluid levels in the front ...

... and rear master cylinder reservoirs – ensure level does not fall past lower mark

that the fuse is of the correct rating and in good condition, and that a spare is available on the machine should the need arise.

At regular intervals remove the battery and check that there is no pale grey sediment deposited at the bottom of the casing. This is caused by sulphation of the plates as a result of re-charging at too high a rate or as a result of the battery being left discharged for long periods. A good battery should have little or no sediment visible and its plates should be straight and pale grey or brown in colour. If sediment deposits are deep enough to reach the bottom of the plates, or if the plates are buckled and have whitish deposits on them, the battery is faulty and must be renewed. Remember that a poor battery will give rise to a large number of minor electrical faults.

If the machine is not in regular use, disconnect the battery and give it a refresher charge every month to six weeks, as described in Chapter 7.

5 Check the brakes

Check that the front and rear brakes work effectively and without binding. Ensure that the rod linkage (rear drum brake) is lubricated and properly adjusted. Check the fluid level in the master cylinder reservoirs, and ensure that there are no fluid leaks. Should topping-up be required, use only the recommended hydraulic fluid.

6 Check the tyres

Check the tyre pressures with a gauge that is known to be accurate. It is worthwhile purchasing a pocket gauge for this purpose because the gauges on garage forecourt airlines are notoriously inaccurate. The pressures should be checked with the tyres **cold**. Even a few miles travelled will warm up the tyres to a point where pressures increase and an inaccurate reading will result.

At the same time as the tyre pressures are checked, examine the tyres themselves. Check them for damage, especially for splitting of the sidewalls. Remove any small stones or other road debris caught between the treads. This is particularly important on the rear tyre, where rapid deflation due to penetration of the inner tube will almost certainly cause total loss of control. When checking the tyres for damage, they should be examined for tread depth. For UK machines, it is vital to keep the tread depth within the legal limits of 1 mm of depth over three-quarters of the tread width around the entire circumference with no sign of bald patches. Many riders consider nearer 2 mm to be the limit for secure roadholding, traction, and braking, especially in adverse weather conditions.

The tyres should be renewed whenever they are found to be damaged or excessively worn (see Chapter 6).

7 Check the controls

Check the throttle and clutch cables and levers, the gear lever and the footrests to ensure that they are adjusted correctly, functioning correctly, and that they are securely fastened. If a bolt is going to work loose, or a cable snap, it is better that it is discovered at this stage with the machine at a standstill, rather than when it is being ridden. Check the operation of the throttle twistgrip and ensure that it snaps shut immediately it is released. If any of the operating cables on the machine appear dry or are stiff in operation, apply a few drops of light machine oil to their exposed sections. If this fails to cure the problem the cable must be removed for thorough lubrication as described under the six-monthly heading.

8 Legal check

Check that all lights, turn signals, horn and speedometer are working correctly to make sure that the machine complies with all legal requirements in this respect. Check also that the headlamp is correctly aimed. Rotate the upper spring-loaded adjusting screw to alter the horizontal alignment; the vertical aim must be aligned so that with the machine standing on its wheels on level ground with the rider (and pillion passenger, if one is regularly carried) seated normally, the dip beam centre (as shown on a wall 25 feet away) must be at the same height from the ground as the centre of the headlamp itself. This is adjusted by another spring-loaded adjusting screw situated at the bottom of the headlamp lens.

Routine maintenance

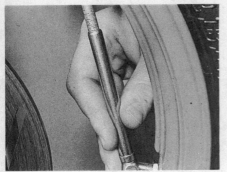
Ensure tyres are run at the correct pressure ...

... and are within the specified wear limits

Headlamp beam adjustment screws; A – horizontal movement, B – vertical movement

Every 300 miles (500 km)

1 Check, adjust and lubricate the final drive chain

The exact interval at which the final drive chain will require lubrication, adjustment and renewal is entirely dependent on the usage to which the machine is put and on the amount of care devoted to chain maintenance. In some cases the chain will require daily lubrication, in other cases it need only be lubricated at weekly intervals. The best rule to follow is that if the chain rollers look dry, then the chain needs lubrication immediately. Do not allow the chain to run dry until the links start to kink, or until traces of reddish-brown deposit can be seen on the sideplates. For ease of reference the full procedure is given here but it must be up to the owner to decide how often the chain needs attention.

Note that two types of chain are available. Standard chains are packed with grease on assembly at the factory and, in addition to normal lubrication, must be removed from the machine at regular intervals so that they can be cleaned thoroughly and re-packed with chain grease as described below. O-ring chains are wider and have small O-rings set between the inner and outer sideplates of each link to seal the grease into the bearings for the life of the chain. These are easily identified by the presence of the O-rings and require a special approach in some aspects of maintenance. Although fitted as standard to all DT and later TZR models, O-ring chains are increasingly popular due to their greatly-increased life and may be encountered on other models.

Cleaning

Although dirt and old lubricant may be washed off the chain while it is in place on the machine, this is not really satisfactory, especially with standard chains where the grease may be washed out of the bearings. To clean the chain thoroughly disconnect it at its connecting link and remove it from the machine. Note that refitting the chain is greatly simplified if a worn out length is temporarily connected to it. As the original chain is pulled off the sprockets, the worn-out chain will follow it and remain in place while the task of cleaning and examination is carried out. On reassembly, the process is repeated, pulling the worn-out chain over the sprockets so that the new chain, or the freshly cleaned and lubricated chain, is pulled easily into place.

To clean the chain, immerse it in a bath containing a mixture of petrol and paraffin and use a stiff-bristled brush to scrub away all the traces of dirt and old lubricant. Take the necessary fire precautions when using this flammable solvent. Swill the chain around to ensure that the solvent penetrates fully into the bushes and rollers and can remove any lubricant which may still be present. When the chain is completely clean, remove it from the bath and hang it up to dry.

Refitting a new, or freshly-lubricated, chain is a potentially messy affair which is greatly simplified by connecting it to the worn-out length of chain used during its removal to pull it around the sprockets. Refit the connecting link, ensuring that the spring clip is fitted with its closed end facing the normal direction of travel of the chain.

Where an O-ring chain is fitted, take care not to lose the O-rings when the connecting link is dismantled, and clean it using kerosene (paraffin) only. Do not use any strong solvents as these may damage the O-rings and never use a steam cleaner or pressure washer. Lubricate the chain thoroughly with the specified oil after cleaning and drying. On reassembly, ensure that the four O-rings are positioned correctly and ensure that there is no space between the connecting link sideplate and spring clip.

Checking for wear

The chain consists of a multitude of small bearing surfaces which will wear rapidly, and expensively, if the chain is not regularly lubricated and adjusted.

Do not omit O-rings (where fitted) from final drive chain joining link

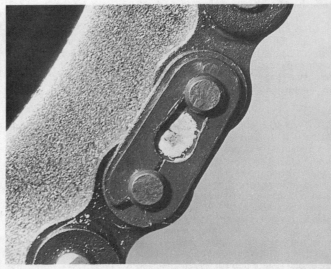
Refit spring clip with closed end in normal direction of chain travel

Routine maintenance

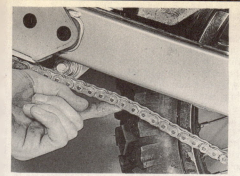
Check chain tension at tightest spot in chain

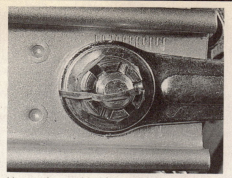
Use marks provided for wheel alignment – TZR model

On DT model use numbered cutouts in snail cams to ensure wheel is aligned

On TZR models, with the chain fully lubricated and correctly adjusted as described below, attempt to pull the chain backwards off the rear sprocket. Yamaha recommend that if the distance between the chain and sprocket is greater than half the pitch of the sprocket teeth, it must be considered worn out and renewed.

The chain fitted to DT models is checked by measuring a 10-link length of the chain (ie mark any one pin, count off 11 pins and measure the distance between the two). This can be done with the chain fitted to the machine. Tension the chain by hand and then measure the distance between the inside of the two pins. Repeat the operation on several different 10 link sections of the chain. If any section of the chain is found to be over 120 mm (4.72 in) it must be considered worn out and renewed.

Renewal

If the chain is found to be worn out, or if any links are kinked or stiff through lack of lubrication, or if damage such as split or missing rollers or side plates is found, the chain must be renewed. This should be done always in conjunction with both sprockets since the running together of new and part-worn components greatly increases the rate of wear of all three items, resulting in even greater expenditure than necessary.

When purchasing a new chain always quote the size, the number of links required and the machine to which the chain is to be fitted. **Standard** chain sizes are given in the Specifications Section of Chapter 1, but note that the gearing may well have been altered on some machines.

Lubrication

On standard type chains, for the purpose of daily or weekly lubrication, one of the many proprietary aerosol-applied chain lubricants can be applied, while the chain is in place on the machine. It should be applied at least once a week, and daily if the machine is used in wet weather conditions. Engine oil can be used for this task, but remember that it is flung off the chain far more easily than grease, thus making the rear end of the machine unnecessarily dirty, and requires more frequent application if it is perform its task adequately. Also remember that surplus oil will eventually find its way on to the tyre, with quite disastrous consequences. While this will serve as a stop-gap measure, it does not reach the inner bearing surfaces of the chain. These can be lubricated correctly only by removing and cleaning the chain as described above and by immersing it in a molten bath of special chain lubricant such as Chainguard or Linklyfe; this should be done at intervals of 500 – 1000 miles for normal road use and more frequently if the machine is used in wet weather or poor conditions. Follow carefully the manufacturer's instructions when using Chainguard or Linklyfe and take great care to swill the chain gently in the molten lubricant to ensure that it penetrates to all bearing surfaces. Wipe off the surplus before refitting the chain to the machine.

Although grease is sealed into the inner bearings of O-ring chains, removing the need for regular immersion in molten lubricant, these chains must still be lubricated at daily or weekly intervals to keep the O-rings supple and to prevent wear between the rollers and sprocket teeth. With the machine supported so that the rear wheel is clear of the ground, revolve the wheel and allow oil to dribble on to the rollers and between the sideplates until the chain is completely oily all along its length. Use only the specified oil for this purpose. **Warning:** some aerosol lubricants contain chemicals which will attack the O-rings causing loss of sealed-in lubricant and premature chain wear; take care to purchase one which is clearly marked as being suitable for O-ring chains.

Adjustment

It is necessary to check the chain tension at regular intervals to compensate for wear. Since this wear does not take place evenly along the length of the chain, tight spots will appear which must be compensated for when adjustment is made. Chain tension is checked with the transmission in neutral and with the machine standing on its wheels. Find the tightest spot on the chain by pushing the machine along and feeling the amount of free play present on the bottom run of the chain, midway between the sprockets, testing along the entire length of the chain. When the tightest spot has been found, measure the total up and down movement available; this should be between the limits given in the Specifications – if not the chain must be adjusted as follows.

TZR model

Remove the split pin and slacken the rear wheel spindle nut by just enough to permit the spindle to be moved. On later models with a disc brake at the rear it will also be necessary to slacken the brake caliper mounting bolt to allow the caliper bracket to move freely. Slacken the chain adjuster locknuts and tighten both drawbolts equally to draw the spindle back by the required amount to take up excess chain slack. To preserve accurate wheel alignment, ensure that the notches stamped in the chain adjusters line up with the same lines stamped on each swinging arm fork end.

When the chain is correctly tensioned, tighten the wheel spindle nut to the specified torque setting and fit a new split pin. Note that on models with a drum brake it will be necessary to apply the rear brake firmly to centralise the brake shoes whilst the nut is tightened. Additionally it might also be necessary to adjust the rear brake and stop lamp switch if the chain has been significantly adjusted; these settings should be checked before taking the machine on the road. Tighten the adjuster locknuts and brake caliper mounting bracket bolt (models with disc brake) and check that the wheel spins freely and that there are no tight spots in the chain.

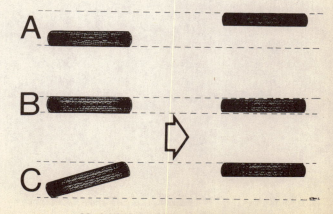
Method of checking wheel alignment

Routine maintenance

DT model

Remove the split pin and slacken the rear wheel spindle nut by just enough to permit the spindle to be moved, then draw the spindle back by rotating the snail cams by the same amount on both sides. Use the numbers stamped in the cams to ensure that the same cam cutout engages with the stopper pin in each swinging arm fork end, thus preserving accurate wheel alignment.

When the chain is correctly tensioned, tighten the spindle nut to the recommended torque setting and fit a new split pin. Finally check that the wheel spins freely and that there are no tight spots in the chain.

All models

On all models, a final check of accurate wheel alignment can be made by laying a plank of wood or drawing a length of string parallel to the machine so that it touches both walls of the rear tyre. Wheel alignment is correct when the plank or string is equidistant from both walls of the front tyre when tested on both sides of the machine, as shown in the accompanying illustration.

Check also that the chain guides and swinging arm pivot guide (where fitted) are straight and unworn. Renew them if they are badly worn by the action of the chain.

Six-monthly, or every 4000 miles (6500 km)

Repeat all previous maintenance tasks, then carry out the following:

1 Check the spark plug

The spark plug supplied as original equipment will prove satisfactory in most operating conditions; alternatives are available to allow for varying altitudes, climatic conditions and the use to which the machine is put. If the spark plug is suspected of being faulty it can be tested only by the substitution of a brand new (not second-hand) plug of the correct make, type, and heat range.

Note that the advice of an authorized Yamaha dealer or similar expert should be sought before the plug heat range is altered from standard. The use of too cold, or hard, a grade of plug will result in fouling and the use of too hot, or soft a grade of plug will result in engine damage due to the excess heat being generated. If the correct grade of plug is fitted, however, it will be possible to use the condition of the spark plug electrodes to diagnose a fault in the engine or to decide whether the engine is operating efficiently or not. The accompanying series of colour photographs will show this clearly.

It is advisable to carry a new spare spark plug on the machine, having first set the electrodes to the correct gap. Whilst spark plugs do not often fail, a new replacement is well worth having if a breakdown does occur. Ensure that the spare is of the correct heat range and type.

The electrode gap can be measured using feeler gauges. If necessary, alter the gap by bending the outer electrode, preferably using a proper electrode tool. **Never** bend the centre electrode, otherwise the porcelain insulator will crack, and may cause damage to the engine if particles break away whilst the engine is running. If the outer electrode is seriously eroded as shown in the photographs, or if the spark plug is heavily fouled, it should be renewed. Renew the spark plug annually regardless of its apparent condition, as it will have passed peak efficiency. Clean the electrodes using a wire brush or a sharp-pointed knife, followed by rubbing a strip of fine emery across the electrodes. If a sand-blaster is used, check carefully that there are no particles of sand trapped inside the plug body to fall into the engine at a later date. For this reason such cleaning methods are no longer recommended; if a plug is that heavily fouled it should be renewed.

Before refitting a spark plug into the cylinder head, coat the threads sparingly with a graphited grease to aid future removal. Use the correct size spanner when tightening the plug, otherwise the spanner may slip and damage the ceramic insulator. The plug should be tightened by hand only at first and then secured with a quarter turn of the spanner so that it seats firmly on its sealing ring. If a torque wrench is available, tighten the plug to the specified torque setting.

Never overtighten a spark plug otherwise there is a risk of stripping the threads from the cylinder head, especially as it is cast in light alloy. A stripped thread can be repaired without having to scrap the cylinder head by using a 'Helicoil' thread insert. This is a low-cost service, operated by a number of dealers.

2 Clean the air filter

On TZR it will be necessary to remove the seat, sidepanels and fuel tank to gain access to the air filter casing. Remove the five screws which secure the filter cover and lift the cover away. The foam element can then be removed.

On DT models, the air filter casing is situated under the left-hand sidepanel. To release the filter element, remove the sidepanel and the three screws which retain the air filter cover and lift the cover away. The foam element can then be withdrawn along with its holder and guide. Remove the wingnut which secures the foam element to the holder and separate them.

Check that the element is not split, hardened or deteriorated through age or severe clogging; renew it if it is damaged in this way. Soak it in a non-flammable or high flash-point solvent such as white spirit (petrol is not recommended because of the fire risk) squeezing it gently to remove all old oil and dirt. Remove excess solvent by squeezing the element between the palms of the hands; wringing it out will damage it. Put the element to one side to alloy any remaining solvent to evaporate.

When it is completely dry, soak the element in clean SAE 10W/30 SE motor oil and gently squeeze out any surplus to leave it only slightly oily to the touch. Refit the element to its holder (DT models) and reassemble with the guide. Apply a light coating of lithium-based grease to both the element and guide seat before refitting. Note, the filter casing on the DT

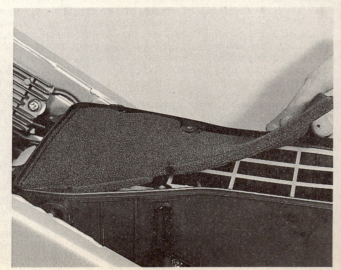

On TZR models remove screws and lift air filter cover away ...   ... foam element can then be removed

Electrode gap check – use a wire type gauge for best results

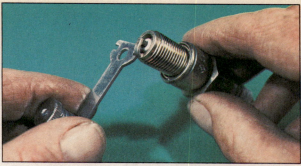
Electrode gap adjustment – bend the side electrode using the correct tool

Normal condition – A brown, tan or grey firing end indicates that the engine is in good condition and that the plug type is correct

Ash deposits – Light brown deposits encrusted on the electrodes and insulator, leading to misfire and hesitation. Caused by excessive amounts of oil in the combustion chamber or poor quality fuel/oil

Carbon fouling – Dry, black sooty deposits leading to misfire and weak spark. Caused by an over-rich fuel/air mixture, faulty choke operation or blocked air filter

Oil fouling – Wet oily deposits leading to misfire and weak spark. Caused by oil leakage past piston rings or valve guides (4-stroke engine), or excess lubricant (2-stroke engine)

Overheating – A blistered white insulator and glazed electrodes. Caused by ignition system fault, incorrect fuel, or cooling system fault

Worn plug – Worn electrodes will cause poor starting in damp or cold weather and will also waste fuel

On DT models remove air filter cover and ...

... withdraw filter assembly

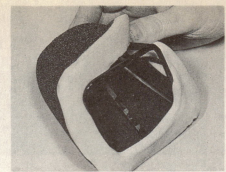
Remove wingnut and washer, then separate foam element from holder

model is fitted with a check hose. If dust or water has accumulated in the hose, the air filter casing must also be cleaned before the filter is refitted.

On reassembly, ensure the element is correctly seated in the filter case (TZR models). On DT models fit the air filter assembly into the casing, making sure its handle is facing outwards. Refit the cover, checking that the sealing O-ring (where fitted) is in good condition and seated correctly. Apply a smear of grease to the cover sealing edges to provide additional protection, then tighten the retaining screws securely. Refit the sidepanel.

Remember that the filter element and cover must be positioned correctly to provide a good seal so that unfiltered air cannot carry dirt into the engine, also that the carburettor is jetted to compensate for the pressure of the air filter; serious damage will occur to the engine through overheating as a result of the weak mixture which will result if the filter is damaged or bypassed in this way. For these reasons the engine should never be run with the air filter removed or disconnected.

This interval is the maximum for filter cleaning; if the machine is used in wet weather or in very dirty or dusty conditions, the filter must be cleaned much more frequently.

3 Check the fuel and engine oil pipes

Give the pipe which connects the fuel tap and carburettor a close visual examination and check for cracks or any signs of leakage. In time, the synthetic rubber pipe will tend to deteriorate, and will eventually leak. Apart from the obvious fire risk, the leaking fuel will affect fuel economy. The pipe will usually split only at the ends; if there is sufficient spare length the damaged portion can be cut off and the pipe refitted. The seal is effected by the interference fit of the pipe on the spigot; although the wire clips are only an additional security measure they should always be refitted correctly and should be renewed if damaged, twisted or no longer effective. If the pipe is to be renewed, always use the correct replacement type and size of neoprene tubing to ensure a good leak-proof fit. Never use natural rubber tubing, as this breaks up when in contact with petrol and will obstruct the carburettor jets, or clear plastic tubing which stiffens to the point of being brittle when in contact with petrol and will produce leaks that are difficult to cure.

Working as described for the fuel line, check the engine oil feed and delivery pipes very carefully. Check that the fitting clamps and wire clips are securely fastened and that the pipes are secured by any clamps or ties provided; ensure that no part of the tubing is chafing on any part of the engine or frame. Any damaged piping must immediately be renewed. If signs of oil leakage or if air bubbles or dirt in the pipes are found, the fault must be cured immediately.

4 Check the carburettor settings and throttle and oil pump cable adjustment

Note: *petrol is extremely flammable, especially when in the form of vapour. Take all precautions to prevent the risk of fire and read the Safety first! section of this manual before starting work.*

If rough running of the engine has developed, some adjustment of the carburettor pilot screw setting and idle speed may be required. If this is the case refer to Chapter 3 for details. Do not make these adjustments unless they are obviously required; there is little to be gained by unwarranted attention to the carburettor. Complete carburettor maintenance by removing the drain plug on the float chamber, turning the petrol on, and allowing a small amount of fuel to drain through, thus flushing any water or dirt from the carburettor. Refit the drain plug securely and switch the petrol off.

Once the carburettor has been checked and reset if necessary, the throttle cable free play can be checked. Open and close the throttle several times, allowing it to snap shut under its own pressure. Ensure that it is able to shut off quickly and fully at all handlebar positions, then check that there is the specified amount of free play measured in terms of twistgrip rotation. Use the adjuster at the twistgrip to achieve the correct setting then open and close the throttle again to settle the cable and check that the adjustment has not altered.

Removing its three retaining screws, withdraw the oil pump cover from the front of the crankcase right-hand cover and check the pump adjustment as follows.

TZR model

The pump pulley has three alignment marks, a round dot and two engraved lines, one each side of the dot. The pump alignment mark is the shorter and lower of the engraved lines. To check the pump synchronization, rotate the twistgrip slightly until all freeplay has been removed from the cable. At this point the alignment mark on the pulley should be directly in line with the pin which protrudes from the centre pivot of the pulley. If this is not the case, slacken the cable adjuster locknut and turn the adjuster, situated on top of the casing, to obtain the required setting. Open and close the throttle a few times, then recheck the setting. Repeat the adjustment procedure as described, then tighten the pump cable adjuster locknut.

DT model

The oil pump setting is controlled by a cable connected via a junction box to the throttle cable. This junction box automatically compensates for wear in the cables and ensures that the oil pump is synchronised accurately with the carburettor at all times. As there is no need for adjustment, it is only possible to check that the pump is operating correctly. To do this, slowly rotate the twistgrip whilst watching the pump pulley closely. The pulley should start to rotate just as the throttle cable takes up its free play and starts to lift the throttle valve. Fully open the throttle. The pulley should rotate smoothly to the fully open position, and should return equally smoothly to its fully closed position when the twistgrip is released and allowed to snap shut under spring pressure.

If the operation of the oil pump is incorrect, the fault is probably in the junction box of the throttle cable. The junction box can be found under the fuel tank, and although no spare parts are available for the box, it can be examined and cleaned by releasing the four screws and removing its cover. Check the box for signs of wear or corrosion; if damaged, the complete throttle cable assembly must be renewed. If all appears to be well, lubricate the junction box components with a synthetic aerosol lubricant such as WD40, do not apply any grease or oil as these could have an adverse effect on the operation of the box. Refit the junction box cover and recheck the operation of the oil pump pulley. If the pulley is still not operating smoothly the throttle cable must be renewed.

Routine maintenance

Adjust oil pump cable until specified marks align at correct throttle position – TZR model

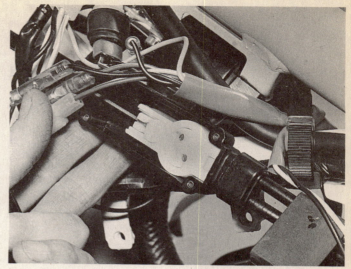

DT model – examine throttle cable junction box if pulley does not operate smoothly

5 Check the transmission oil level

With the machine placed on level ground, start the engine and allow it to warm up to normal operating temperature. Stop the engine, wait a minute for the transmission oil level to settle, and then check the level via the sightglass situated in the right-hand outer casing. With the machine standing absolutely upright, the level should be between the maximum and minimum marks on the side of the glass. If the oil level is too low remove the filler cap and add sufficient oil to raise it to the required level. Use only a good quality oil of the specified grade. Inspect the filler cap O-ring and renew it if necessary. Refit the filler cap.

6 Adjust the clutch

The clutch is adjusted correctly when there is 2 – 3 mm (0.08 – 0.20 in) of freeplay between the lever butt end and its mounting bracket. If adjustment is required, use the handlebar end adjuster on the lever mounting bracket. If there is insufficient range in the adjuster, screw it fully inwards and reset the lower adjuster, mounted on the left-hand side of the barrel, to take up the excess freeplay. Fine adjustment can then be made at the handlebar lever adjuster. If both adjusters are fully extended and there is still excess freeplay, the clutch release mechanism must be adjusted. To gain access to the clutch, it will be necessary to carry out the following, referring to the relevant sections of Routine maintenance and Chapter 1 for details.

Transmission oil should be between marks on sightglass

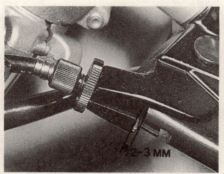

Adjust clutch cable to obtain correct amount of freeplay using handlebar and ...

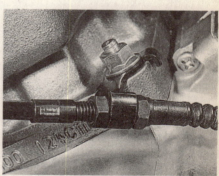

... lower adjuster

If necessary, adjust clutch release mechanism (see text) ...

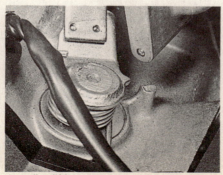

... to align lifting arm pointer with raised mark – DT model

On TZR model, pointer is on end of lifting arm

Routine maintenance

(a) Slacken fully both cable adjusters
(b) Remove the oil pump cover and disconnect the oil pump cable and oil pipes
(c) Drain the transmission oil
(d) Drain the cooling system
(e) Disconnect the radiator hose from the side cover
(f) Remove the kickstart lever
(g) Remove the right-hand crankcase cover

The clutch release mechanism adjuster is situated in the centre of the pressure plate and is adjusted as follows:

Slacken the pushrod adjuster locknut and push the clutch lifting arm as far forward as possible. Then slowly screw in the pushrod adjusting screw whilst closely watching the lifting arm, the arm is correctly adjusted when the pointer on the lifting arm aligns with the raised mark on the crankcase, and tighten the locknut. Complete adjustment by readjusting the cable as described above.

If the required amount of free play still cannot be obtained renew the clutch cable. If the problem persists, the clutch assembly must be dismantled and examined as described in Chapter 1.

7 Check and adjust the brakes

Front brake

The hydraulic brake requires no regular adjustments; pad wear is compensated for by the automatic entry of more fluid into the system from the fluid reservoir. All that is necessary is to maintain a regular check on the fluid level and the degree of pad wear.

To check the fluid level, turn the handlebars until the reservoir is horizontal (front brake only) and check that the fluid level, as seen through the reservoir body, is not below the lower level mark. Remember that while the fluid level will fall steadily as the pad friction material is used up, if the level falls below the lower level mark there is a risk of air entering the system; it is therefore sufficient to maintain the fluid level above the lower level mark, by topping-up if necessary. Do not top up to the higher level mark (where shown) unless this is necessary after new pads have been fitted. If topping up is necessary, wipe any dirt off the reservoir, remove the retaining screws and lift away the reservoir cover and withdraw the diaphragm. Use only good quality brake fluid of the recommended type and ensure that it comes from a freshly opened sealed container. Brake fluid is hygroscopic, which means that it absorbs moisture from the air, therefore old fluid may have become contaminated to such an extent that its boiling point has been lowered to an unsafe level. Remember also that brake fluid is an excellent paint stripper and will attack plastic components; wash away any spilled fluid immediately with copious quantities of water. When the level is correct, clean and dry the diaphragm, fold it into its compressed state and fit it to the reservoir. Refit the reservoir cover and tighten securely, but do not overtighten, the retaining screws.

To check the degree of pad wear on TZR models, look closely at the pads from behind the caliper. Wear limit marks are provided on the corners of the metal backing of the pads. If these markers are making or are close to making contact with the disc, the pads must be renewed as a set.

On the DT model caliper, there is a removable rubber plug for inspecting the pads. The wear indicator is in the form of a groove running down the centre of the friction material. If the pads have worn to the point where this groove is exposed, the pads must be renewed as a set.

To remove the pads on earlier TZR models (2RK and 3PC1), remove both caliper mounting bolts and manoeuvre the brake caliper off the disc. Whilst supporting the caliper so that it does not hang by its hydraulic hose, slacken and remove the two brake pad retaining bolts. The pads should now be free and can be slid out for examination or renewal; note the positions of the anti-squeal shims.

On the later TZR models (3PC2 and 3PC3), the brake pads can be removed from the caliper whilst the caliper is fitted to the machine. Lever off the cover from the top of the caliper to reveal the pads and pad spring. Remove the two R-clips from the pad retaining pins and withdraw the pins and pad spring. The brake pads can now be lifted out of the caliper for examination or renewal. Note the positions of the pad spring and anti-squeal shims as a guide to reassembly.

On DT models, remove the disc cover retaining screws and remove both the disc and caliper covers. Slacken and remove the caliper body retaining bolt, the caliper can then be rotated in an anticlockwise direction to reveal the brake pads. The pads can then be removed from

Ensure that diaphragm is correctly positioned before refitting brake fluid reservoir cover

On TZR model brake caliper must be viewed from rear to check pad wear

On DT model remove rubber plug to inspect brake pads

To remove brake pads, remove caliper mounting bolts (TZR 2RK and 3PC1 models) ...

... and slacken and remove pad retaining bolts

Remove pads from caliper noting correct positions of anti-squeal shims

Routine maintenance

On DT models remove disc and caliper covers to gain access to front caliper

On TZR models ensure anti-rattle spring is correctly installed in caliper ...

... and also that anti-squeal shims are correctly positioned on back of brake pads

the mounting bracket along with the pad springs. Make a note of how the springs were positioned as a guide to reassembly.

If the pads are worn to the limit marks, fouled with oil or grease, or heavily scored or damaged by dirt and debris, they must be renewed as a set; there is no satisfactory way of degreasing friction material. If the pads can be used again, clean them carefully using a fine wire brush that is completely free of oil or grease. Remove all traces of road dirt and corrosion, then use a pointed instrument to clean out the groove in the friction material (TZR models) and to dig out any embedded particles of foreign matter. Any areas of glazing may be removed using emery cloth. Measure the thickness of the pad friction material. If either pad has worn to or beyond the service limit, given in the Specifications, the pads must be renewed as a set.

On reassembly, if new pads are to be fitted, the caliper piston(s) must now be pushed back as far as possible into the caliper bore(s) to provide the clearance necessary to accommodate the unworn pads. It should be possible to do this with hand pressure alone. If any undue stiffness is encountered the caliper assembly should be dismantled for examination as described in Chapter 6. While pushing the piston(s) back, maintain a careful watch on the fluid level in the handlebar reservoir. If the reservoir has been overfilled, the surplus fluid will prevent the pistons returning fully and must be removed by soaking it up with a clean cloth. Take care to prevent fluid spillage. Apply a thin smear of caliper grease to the edges and rear surface of the pads and, on TZR models, to the pad retaining pins. Take care to apply caliper grease to the metal backing of the pads only and not to allow any grease to contaminate the friction material.

Fit the pads by reversing the removal sequence. Note that Yamaha recommend that the shims (where fitted) and pad springs should be renewed each time new pads are fitted. On TZR models ensure that the pads locate correctly against the anti-squeal shims and that the anti-rattle spring is correctly installed in the caliper. On later models (3PC2 and 3PC3) note that the anti-rattle spring must be fitted with the stamped arrow pointing in the direction of wheel rotation. Place the pads in the caliper and refit the brake pad retaining bolts, tightening them to the specified torque setting (2RK and 3PC1 models), or the brake pad retaining pins and R-pins (3PC2 and 3PC3 models). On the earlier TZR models (2RK and 3PC1), refit the caliper assembly to the machine ensuring that the pads engage correctly on the disc and that the hose is not twisted, then refit the mounting bolts and tighten them to the specified torque setting.

On DT models note that the rounded side of the pads should face away from the disc. Check that the two pad springs are located correctly and swing the caliper body back down into position. Refit the caliper body retaining bolts having first applied a smear of caliper grease to its shank. Tighten the retaining bolt to the specified torque setting and refit the disc and caliper covers.

Apply the brake lever gently and repeatedly to bring the pads firmly into contact with the disc until full brake pressure is restored. Be careful to watch the fluid level in the reservoir; if the pads have been re-used it will suffice to keep the level above the lower level mark, by topping-up if necessary, but if new pads have been fitted the level should be slightly higher to allow for the pads to bed in. Refit the reservoir cover, and diaphragm as described above and adjust the brake lever freeplay.

Before taking the machine out on the road, be careful to check for fluid leaks from the system, and that the front brake is working correctly. Remember also that new pads, and to a lesser extent, cleaned pads will require a bedding-in period before they will function at peak efficiency. Where new pads are fitted use the brake gently but firmly for the first 50-100 miles to enable the pads to bed in fully.

Rear brake

The rear brake system fitted to the DT model is basically similar to the front and the remarks made can be applied in most respects. To gain access to the rear caliper it will first be necessary to remove the caliper guard plate which is bolted to the swinging arm. Remove the caliper body retaining bolt and pivot the caliper upwards to gain access to the pads. Refer to the details given for the front brake above, noting that the brake pad on the piston side of the caliper (outside) is fitted with an anti-squeal shim. This should be renewed each time a new set of pads is fitted. Make sure that the shim is fitted correctly and to the correct brake pad.

Early TZR models (2RK, 3PC1) are fitted with a drum brake whereas later models (3PC2 and 3PC3) use a disc brake.

The rear brake caliper is basically the same as the front caliper on early TZR models (2RK and 3PC1); the remarks made previously can therefore be applied.

The rear drum brake fitted to the early models should be overhauled as follows:

Check the condition of the brake operating linkage, ensuring that all components are in good condition, properly lubricated and securely fastened. Dismantle the rear brake pedal pivot and grease it at regular intervals. Note that the operating mechanism is at its most efficient when, with the brake correctly adjusted and applied fully, the angle between the cable or rod and the operating arm on the brake backplate does not exceed 90°. This can be adjusted by removing the operating arm from the brake camshaft and rotating it by one or two splines until the angle is correct. Ensure that all components are correctly secured on reassembly.

Drum brake shoe wear is checked by applying the brake firmly and looking at the wear indicator marks on the backplate. If the indicator pointer on the camshaft is aligned with, or has moved beyond, the fixed index mark cast on the backplate, the shoes are worn out and must be renewed as a pair. This involves the removal of the wheel from the machine as described in Chapter 6 so that the brake components can then be dismantled, cleaned, checked for wear, and reassembled following the instructions given in the same Chapter. It is important that moving parts such as the brake camshaft are lubricated with a smear of high melting-point grease on reassembly (see two-yearly heading).

To adjust the rear brake, support the machine so that the rear wheel is clear of the ground, spin the wheel and tighten the adjusting nut until a rubbing sound is heard as the shoes begin to contact the drum, then slacken the nut by one or two turns until the sound ceases. Spin the wheel and apply the brake hard to settle the components. Check that the adjustment has not altered and tighten securely all disturbed fasteners. This should approximate the specified setting of free play before the brake begins to engage the drum (see Specifications). Switch on the ignition and check that the stop lamp lights just as all free play has been taken up and the brake is beginning to engage. This is adjusted by holding steady the stop lamp rear switch body and rotating the plastic sleeve nut to raise or lower the switch as necessary; do not allow the body to rotate or its terminal will be damaged. Check the switch setting whenever the rear brake is adjusted.

Complete brake maintenance by checking that the wheels are free to rotate easily, and then lubricate all linkage pivots. To prevent the risk of oil finding its way on to the tyres or the brake shoe linings do not oil the brake components excessively; a few drops of oil at each point will suffice. Lubricate the stop lamp switch with a water dispersant fluid such as WD40.

DT models (rear brake) – remove guard to gain access to caliper

Release caliper body retaining bolt ...

... and rotate caliper away to reveal pads – note shim fitted to outer pad

Ensure anti-rattle springs are correctly positioned before refitting pads

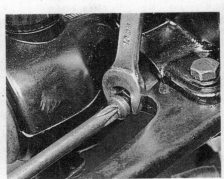
Adjust front brake lever free play using screw and locknut

On drum brake models, adjust rear brake at end of operating rod

Brake lever and pedal adjustments – all models

To complete brake maintenance it will be necessary to check that the brake lever and pedal are correctly adjusted.

The front brake lever is adjusted by a screw and locknut situated on the handlebar lever. When properly adjusted there should be 2 – 5 mm (0.08 – 0.20 in) freeplay, measured at the tip of the brake lever. If not slacken the locknut and adjust the screw until the correct amount of freeplay is obtained. Once correctly adjusted, tighten the locknut and check the operation of the brake before taking the machine on the road.

On the rear drum brake it will first be necessary to check that the pedal is set at the correct height. The tip of the brake pedal should be 41 mm (1.6 in) below the top of the footrest. If not, the pedal height can be altered by slackening its locknut and moving the adjuster situated on the footrest bracket. Once the pedal height is known to be correct the brake pedal free play can be checked. There should be 20 – 30 mm (0.8 – 1.2 in) of free play, measured at the brake pedal tip before the brake shoes come in contact with the drum. This is adjusted by turning the nut on the end of the brake rod. Finally it will be necessary to check the stop lamp switch setting whenever any adjustment is carried out on the brake pedal. The switch can be adjusted by turning its mounting nut whilst holding the body of the switch to prevent it from rotating. The stop lamp should operate just as the brake shoes make contact with the drum. Check the operation of the rear brake and stop lamp switch thoroughly before taking the machine on the road

On the rear disc brake check that the rear brake pedal is set to the correct height as shown in the Specifications. If not, the brake pedal can be adjusted using the two nuts on the bottom of the master cylinder operating rod. On TZR models the top nut is the locknut and the bottom nut the adjuster, whereas on DT models the bottom nut is the locknut and the pedal is adjusted using the top nut. Slacken the locknut and move the adjuster until the correct pedal height is obtained. Once the pedal height is set, make a visual check to see that the adjustment mechanism is correct. On TZR models the threaded adjuster should protrude 3 – 5 mm (0.12 – 0.20 in) below the lower nut, whereas on DT models it should protrude at least 2 mm (0.08 in) out of the lower adjusting nut and be at least 2 mm (0.08 in) away from the brake pedal. Once the adjuster has been checked tighten the locknut and check that the rear wheel rotates freely. Check that the stop lamp operates correctly and adjust its position if necessary; use the information in the drum brake section for details. Thoroughly check the operation of the rear brake before taking the machine on the road.

8 Check the wheels and wheel bearings

Position the machine on a suitable stand and raise the wheel to be examined clear of the ground, using blocks positioned beneath the engine. Check the wheel spins freely; if necessary, slacken the brake adjuster or push back the brake pads, and in the case of the rear wheel, detach the final drive chain.

DT model

Examine the rim for serious corrosion or impact damage. Slight deformities can often be corrected by adjusting spoke tension. Serious damage and corrosion will necessitate renewal, which is best left to an expert. A light alloy rim will prove more corrosion resistant.

Place a wire pointer close to the rim and rotate the wheel to check it for runout in the radial and axial planes. If the rim is more than 2.0 mm (0.08 in) out of true in either plane, check spoke tension by tapping them with a screwdriver. A loose spoke will sound quite different to those around it. Worn bearings will also cause rim runout.

Adjust spoke tension by turning the square-headed nipples with the appropriate spoke key which can be purchased from a dealer. With the spokes evenly tensioned, remaining distortion can be pulled out by tightening the spokes on one side of the wheel and slackening those directly opposite. This will pull the rim across whilst maintaining spoke tension.

More than slight adjustment will cause the spoke ends to protrude through the nipple and chafe the inner tube, causing a puncture. Remove the tyre and tube and file off the protruding ends. The rim band protects the tube against chafing – check it is in good condition before fitting.

Check spoke tension and general wheel condition regularly. Frequent cleaning will help prevent corrosion. Replace a broken spoke immediately because the load taken by it will be transferred to adjacent spokes which may fail in turn.

TZR model

Carefully check the complete wheel for cracks and chipping, particularly at the spoke roots and the edge of the rim. As a general rule a damaged wheel must be renewed as cracks will cause stress points which may lead to sudden failure under heavy load. Small nicks may be radiused carefully with a fine file and emery paper (No 600 – No 1000) to relieve the stress. Note this will destroy the painted finish of the wheel and the wheel will thus require touching in with a suitable paint. If there is any doubt as to the condition of a wheel, advice should be sought from a reputable dealer or specialist repairer.

Check the axial runout at the rim by spinning the wheel and placing a fixed pointer close to the rim edge. If the maximum runout is greater than the specified limit, the manufacturer recommends that the wheel be renewed. This is, however, a counsel of perfection; a runout somewhat greater than this can probably be accommodated without noticeable effect on steering. No means is available for straightening a warped wheel without resorting to the expense of having the wheel skimmed on both faces. If warpage was caused by impact during an accident, the safest measure is to renew the wheel complete. Worn wheel bearings may cause rim runout. These should be renewed.

All models

An out-of-balance wheel will produce a hammering effect through the steering at high speed. Spin the wheel several times. A well balanced wheel will come to rest in any position. One that comes to rest in the same position will have its heaviest part downward and weights must be added to a point diametrically opposite until balance is achieved. Where the tyre has a balance mark on its sidewall (usually a coloured spot), check it is in line with the valve.

To check the wheel bearings, grasp each wheel firmly at the top and bottom and attempt to rock it from side to side; any free play indicates worn bearings which must be renewed as described in Chapter 6.

9 Check the steering and suspension

Raise the front wheel off of the ground by placing the machine on a suitable stand and securing the machine firmly.

Check the steering head bearing adjustment by grasping the bottom of both fork lower legs, then pulling and pushing in a fore and aft direction; any free play should be felt between the fork bottom yoke and the frame head lug. Check for overtightened bearings by placing the forks in the straightahead position and tapping lightly on one handlebar end; the forks should fall away smoothly and easily to the opposite lock, taking into account the effect of cables and wiring, with no traces of notchiness. If necessary, adjust the steering head bearings using the information in Chapter 5.

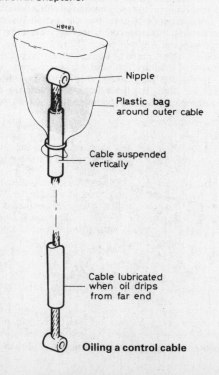

Oiling a control cable

Ensure that the front forks work smoothly and progressively by pumping them up and down whilst the front brake is held on. Any faults revealed by this check should be investigated immediately; refer to Chapter 5, particularly if the oil seals are leaking. On TZR models inspect the stanchions, looking for signs of chips or other damage, then lift the dust excluder at the top of each fork lower leg and wipe away any dirt from its sealing lips and above the fork oil seal. Pack grease above the seal and refit the dust excluder. Carry out the same procedure on DT models, having first released its clamp and pulled down the fork gaiter. Note that damage to the stanchion is unlikely on this model due to the protection given by the gaiters.

Check that all the rear suspension pivot bolts are secure and that the pivots operate smoothly. Check for free play in the swinging arm by grasping its forked ends and pushing and pulling it horizontally. If free play is found the swinging arm must be removed for further examination, as described in Chapter 5. On DT models check the suspension linkage relay arm pivots for free play. If free play is found, the linkage must be overhauled as described in Chapter 5. If all is well, lubricate the swinging arm pivot and linkage bushes with a lithium-based grease. DT models are fitted with grease nipples for this purpose and the grease can be applied with a suitable grease gun. On TZR models, however, it will be necessary to remove the swinging arm pivot bolt to apply the grease as no grease nipples are fitted. This technique can also be used on the DT model if a suitable grease gun is not available.

The rear suspension unit preload can be adjusted to suit the rider's weight, riding style and road condition; refer to Chapter 5, Section 15 for details of adjustment.

10 Check the cooling system

The cooling system should be checked for signs of leakage or damage and any suspect hoses renewed. Refer to Chapter 2 for further information.

11 Lubricate the control and instrument drive cables

Check the outer cables for signs of damage, then examine the exposed portions of inner cables. Any signs of kinking or fraying will indicate that renewal is required. To obtain maximum life and reliability from the cables they should be thoroughly lubricated. To do the job properly and quickly use one of the hydraulic cable oilers available from most motorcycle shops. Free one end of the cable and assemble the cable oiler as described by the manufacturer's instructions. Operate the oiler until oil emerges from the lower end, indicating that the cable is lubricated throughout its length. This process will expel any dirt or moisture and will prevent its subsequent ingress.

If a cable oiler is not available, an alternative is to remove the cable from the machine. Hang the cable upright and make up a small funnel arrangement using plasticine or by tapping a plastic bag around the upper end. Fill the funnel with oil and leave it overnight to drain through. Note that where nylon-lined cables are fitted, they should be used dry or lubricated with a silicone-based lubricant suitable for this application. On no account use ordinary engine oil because this will cause the liner to swell, pinching the cable. On DT125 models only, remember to inspect and to lubricate, if necessary, the throttle/oil pump cable junction box as described under item 4 of this Section.

Similarly, to ensure the correct and accurate operation of the speedometer and tachometer, the drive cables must be removed from the machine as described in Section 18 of Chapter 5 and greased. Remember to grease all but the upper six inches of the cable to prevent damage to the instrument head.

12 Check and lubricate the side stand pivot

It is essential that the smooth, safe, operation of the side stand is ensured by regular lubrication, which will serve two purposes. The first, and most important, is that a layer of grease will prevent the onset of corrosion which would cause the controls to become stiff and jerky in operation, and the second is that lubrication will minimise the wear that will inevitably occur on any bearing surface.

Dismantle, clean and grease the side stand pivot as described in Section 16 of Chapter 5.

Check that the side stand switch is functioning correctly and lubricate it with a water dispersant fluid such as WD40.

Lubricate sidestand pivot and apply a water-dispersant fluid to switch

Transmission oil drain plug is situated at bottom of right-hand crankcase cover

Annually, or every 8000 miles (13 000 km)

The annual maintenance operation should be regarded as a minor overhaul. Carry out all the applicable tasks listed in the previous mileage/time service headings, then complete the following:

1 Change the transmission oil

Oil changes are much quicker and more efficient if the machine is taken for a journey long enough to warm the oil up to normal operating temperature so that it is thin and is holding any impurities in suspension.

Position a suitable container of at least 1 litre (1.8 pint) beneath the engine right-hand side and remove the drain plug from the bottom of the right-hand crankcase cover. Remove the oil filler plug from the top of the casing. Whilst waiting for the oil to drain, examine the plug sealing washer and renew if damaged. Tilt the machine slightly to the right to allow the oil to drain fully.

On completion of draining refit the drain plug and sealing washer, tightening it to the torque setting given in the Specifications. Fill the gearbox with the specified amount of SAE 10W/30 SE engine oil. Check the oil level as described in the six-monthly/4000 mile service interval and refit the filler plug. Check carefully for any signs of oil leaks.

2 Renew the brake fluid

It is necessary to renew the brake fluid at this interval to preserve maximum brake efficiency by ensuring that the fluid has not been contaminated and deteriorated to an unsafe degree.

Before starting work, obtain a new, full can of the recommended hydraulic fluid and read carefully the Section on brake bleeding in Chapter 6. Prepare the clear plastic tube and glass jar in the same way as for bleeding the hydraulic system, open the bleed nipple by unscrewing it ¼ – ½ a turn with a spanner and apply the front brake lever or rear brake pedal (as applicable) gently and repeatedly. This will pump out the old fluid. **Keep the master cylinder reservoir topped up at all times,** otherwise air may enter the system and greatly lengthen the operation. The old brake fluid is invariably much darker in colour than the new, making it easier to see when it is pumped out and the new fluid has completely replaced it.

When the new fluid appears in the clean plastic tubing completely uncontaminated by traces of old fluid, close the bleed nipple, remove the plastic tubing and replace the rubber cap on the nipple. Top the master cylinder reservoir up to above the lower level mark, unless the brake pads have been renewed in which case slightly more fluid should be added. Clean and dry the rubber diaphragm, fold it into its compressed state and refit the diaphragm and reservoir cover, tightening securely the retaining screws (where applicable).

Wash off any surplus fluid and check that the brake is operating correctly before taking the machine out on the road.

3 Change the front fork oil

Note: *On 3RN8 models there is no oil drain plug in the fork lower leg. To change the fork oil, remove the forks from the yokes (see Chapter 4, Section 2) and then remove the top bolt, spacer, spring seat and spring, invert each fork and allow the oil to drain (see Chapter 5, Section 3, paras 15 and 16).*

Place a sheet of cardboard against the wheel to keep oil off the brake or tyre, place a suitable container under the fork leg and remove the drain plug which is a small screw at the side of the fork lower leg.

Depress the forks several times to expel as much oil as possible, then repeat the process on the remaining leg. Leave the machine for a few minutes to allow any residual oil to drain to the bottom, then pump the forks again to remove it.

Examine the drain plug sealing washers and renew them if necessary. Refit the drain plugs and place the machine on a suitable stand so the front wheel is clear of the ground and the forks are fully extended. Remove the top bolt (DT models) or top plug and circlip (TZR models) to gain access to the inside of the stanchion. Remove the fork spring and on DT models, also remove the spring seat and spacer, and fill each fork leg with the specified amount of the recommended oil. Check the fork oil level by making up a dipstick, cut to the dimension given in the Specifications, and passing this down the stanchion; note that the forks should be fully compressed for this check. Note the measurement, which should be taken in the centre of the fork, and make any adjustments necessary to the oil quantity. Refit the fork spring, spring seat and spacer, followed by the top plug and circlip or top bolt. Move the machine off of its stand, apply the front brake and pump the forks up and down until the damping effect can be felt to be fully restored.

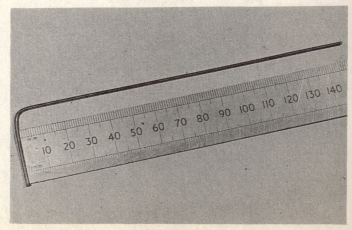

Fork oil level dipstick can be fabricated from a length of wire as shown

4 Renew the spark plug

The spark plug should be renewed at this interval, regardless of its apparent condition, as it will have passed peak efficiency. Check that the new plug is of the correct type and heat range and that it is gapped correctly before it is fitted. Coat its threads with a graphited grease prior to installation, and tighten to the specified torque setting.

5 Clean the fuel filters

Note: *petrol is extremely flammable, especially when in the form of vapour. Take all precautions to prevent the risk of fire and read the Safety first! section of this manual before starting work.*

Remove the fuel tap from the fuel tank as described in Section 3 of Chapter 3. Remove both the tubular fuel filters from the tap and clean them carefully in new petrol. Refer to Chapter 3, Section 3 and refit the filters and tap; check for fuel leakage before taking the machine out on the road.

On the flat slide type carburettor fitted to later DT models, there is also a fuel filter fitted to the carburettor body. This filter is situated on the fuel inlet side and can be removed from the carburettor once the fuel tap has been turned off and the fuel pipe has been disconnected from its carburettor union. Remove the filter, clean it in new petrol and refit it to the carburettor. Connect the fuel pipe and turn on the fuel tap. Thoroughly examine the fuel tap and pipe for leaks before taking the machine on the road.

Two-yearly, or every 16 000 miles (25 000 km)

Complete the tasks listed under the previous headings, then carry out the following:

1 Grease the steering head bearings

Referring to the relevant Sections of Chapter 5, dismantle the steering head, clean all components and check them for wear, renewing any worn items. Reassemble, packing the steering head bearings with fresh grease.

2 Renew the coolant

To minimise the build-up of deposits in the cooling system and to ensure maximum protection against its freezing, the coolant should be drained completely, the system should be flushed out and checked for leaks or damage and new coolant mixed for refilling. Refer to Chapter 2 for details.

3 Grease wheel bearings, speedometer drive and brake camshaft

To prolong their life as much as possible, the front and rear wheels should be removed from the machine so that their bearings can be driven out, cleaned, checked for wear and renewed if necessary, then repacked with the recommended grease and refitted. At the same time the speedometer drive should be cleaned and packed with grease, and on drum brakes only, the brake camshaft should be removed, cleaned and lightly greased. Refer to the relevant Sections of Chapter 6.

Additional routine maintenance

Certain tasks do not fall under the previous mileage/time headings as they concern items which deteriorate with age, whether the machine is used a great deal, or hardly at all.

1 Cleaning the machine

Keeping the motorcycle clean should be considered as an important part of the routine maintenance, to be carried out whenever the need arises. A machine cleaned regularly will not only succumb less speedily to the inevitable corrosion of external surfaces, and hence maintain its market value, but will be far more approachable when the time comes for maintenance or service work. Furthermore, loose or failing components are more readily spotted when not partially obscured by a mantle of road grime and oil.

Surface dirt should be removed using a sponge and warm, soapy water, the latter being applied copiously to remove the particles of grit which might otherwise cause damage to the paintwork and polished surfaces.

Oil and grease are removed most easily by the application of a cleaning solvent such as 'Gunk' or 'Jizer'. The solvent should be applied when the parts are still dry and worked in with a stiff brush. Large quantities of water should be used when rinsing off, taking care that water does not enter the carburettor, air filter or electrics.

Application of a wax polish to the cycle parts and a good chrome cleaner to the chrome parts will also give a good finish. Always wipe the machine down if used in the wet, and make sure the chain is well oiled. There is less chance of water getting into control cables if they are regularly lubricated, which will prevent stiffness of action.

2 Decarbonisation

The oily nature of any two-stroke engine's exhaust leads to layers of carbon being deposited in the combustion chamber and exhaust system. If not removed at regular intervals these deposits will built up to the point where the machine's performance and economy are significantly reduced.

It is very difficult to give a precise interval for decarbonisation as so many different factors have to be taken into account. For example if maintenance is neglected so that deposits build up at a faster rate through inefficient combustion, or if the wrong type of engine oil is used. Furthermore, the rider's driving style must be taken into account; any machine that is used principally for fast riding or for long journeys on open roads will not require decarbonisation as often as a machine which is used for short, low-speed commuting trips. Some machines will require decarbonisation, therefore, once a year while others will run satisfactorily for more than twice as long before attention is required. As an initial starting point is suggested that this task be carried out once a year until sufficient experience is gained for the interval to be shortened or extended as necessary. Note that it may not be necessary to decarbonise the exhaust as frequently as the engine, or vice versa, so the two can be treated as separate items, with their own schedules.

Cylinder head and barrel

Refer to Chapter 1 and remove the cylinder head and barrel. It is necessary to remove all carbon from the head, barrel and piston crown whilst avoiding removal of the metal surface on which it is deposited. Take care when dealing with the soft alloy head and piston. Never use a steel scraper or screwdriver. A hardwood, brass or aluminium scraper is ideal as these are harder than the carbon but not harder than the underlying metal. With the bulk of carbon removed, use a brass wire brush. Finish the head and piston with metal polish; a polished surface will slow the subsequent build-up of carbon. Clean out the barrel ports to prevent the restriction of gas flow and clean the YPVS components. Remove all debris by washing each component in paraffin whilst observing the necessary fire precautions. Renew the piston rings, if necessary, on reassembly.

Exhaust system

First check that the exhaust system is securely fastened and that there are no exhaust leaks. Renew the exhaust port gasket and, on DT models only, the rubber seal between the two parts of the system, if leaks are found at either of these points. Some idea of the condition of the exhaust can be gained by looking at the tailpipe. If the mixture and oil pump settings are correct, and the machine is ridden normally, there should be a thin film of sooty black carbon with a slight trace of oiliness. If the machine is ridden hard, the deposits will tend to be a lighter colour, almost grey, and there will be no traces of oil. If any of the above is found, and the spark plug electrode colour, carburettor settings, air filter condition, and the oil pump settings are known to be satisfactory from the previous Routine Maintenance operations, then the carbon deposits inside the exhaust system will be kept to a minimum and will take a long time to build up to the point where a full decarbonising operation is necessary. On the other hand, if the rear of the exhaust is excessively oily and the carbon deposits rather thicker than those described above, then either one of the settings mentioned above is incorrect, causing the engine to run inefficiently so that it produces too much waste in the form of carbon, or the machine is being ridden too slowly which produces the same symptoms. If the engine settings are known to be satisfactory thanks to the previous Routine Maintenance operations and the carbon build-up is caught at an early stage, then the simplest and most satisfying method of cleaning the exhaust is to take the machine on a good hard run until the exhaust is too hot to touch and the excessive smoke produced by the dispersal of the carbon/oil build-up has disappeared. This is a well-known trick employed by many mechanics to restore lost power, and can have quite dramatic results. If the build-up has been allowed to develop too much, however, the complete exhaust must be removed from the machine and decarbonised using one of the methods described below.

Three possible methods of cleaning the system exist; the first, which will only be really effective if the deposits are very oily, consists of flushing the system with a petrol/paraffin mixture. It will be evident that great care must be taken to prevent the risk of fire when using this method, and that the system must be hung overnight in such a way that all the mixture will drain away. It will be necessary to use a suitable scraping tool to remove any hardened deposits from the front length of the exhaust pipe and from the exhaust port of the cylinder barrel.

The second method will involve the use of a welding torch or powerful blow lamp to burn off the deposits. This usually results in the production of a great deal of smoke and fumes and so must be carried out in a well-ventilated area. It also requires some considerable skill in the use of welding equipment if it is to be fully effective, and if personal injury or damage to the exhaust system is to be avoided. Another drawback is that the excessive heat created will destroy the painted finish of the system, and that repainting will be necessary. With severely blocked systems, it may even be necessary to cut the exhaust open so that the flame can be applied to the blocked section, and the exhaust welded up afterwards. In short this method, while quite effective, should only be employed by a person with welding equipment and the necessary degree of skill in its use. The best solution would be for the owner to remove the exhaust from the machine and to take it to a local dealer or similar expert for the work to be carried out.

The third possible method of exhaust cleaning is to use a solution such as caustic soda to dissolve the deposits. While this is a lengthy and time-consuming operation, it is the simplest and most effective method that can be used by the average owner, and is therefore described in detail in the following paragraphs. To clean the silencer casing and exhaust pipe assembly, remove the system from the machine and suspend it from its silencer end. Block up the end of the exhaust pipe with a cork or wooden bung. If wood is used, allow an outside projection of three or four inches with which to grasp the bung for removal.

Bear in mind that it is very important to take great care when using caustic soda as it is a very dangerous chemical. Always wear protective clothing, this must include proper eye protection. If the solution does come into contact with the eyes or skin it must be washed clear immediately with clean, fresh, running water. In the case of an eye becoming contaminated, seek expert medical advice immediately. Also, the solution must not be allowed to come into contact with aluminium alloy – especially at the above recommended strength – caustic soda reacts violently with aluminium and will cause severe damage to the component.

The mixture used is a ratio of 3 lbs caustic soda to a gallon of fresh water. This is the strongest solution ever likely to be required. Obviously the weaker the mixture the longer the time required for the carbon to be dissolved. Note, whilst mixing the solution, that the caustic soda should be added to the water gradually, whilst stirring. Never pour water into a container of caustic soda powder or crystals; this will cause a violent reaction to take place which will result in great danger to one's person.

Commence the cleaning operation by pouring the solution into the system until it is quite full. Do not plug the open end of the system. The solution should not be left overnight for its dissolving action to take place. Note that the solution will continue to give off noxious fumes throughout its dissolving process; the system must therefore be placed in a well ventilated area. After the required time has passed, carefully pour out the solution and flush the system through with clean, fresh water. The cleaning operation is now complete.

3 Renew the brake caliper seals and master cylinder seals

Yamaha recommend that the brake caliper and master cylinder seals should be renewed at two-yearly intervals, as a safety precaution against their sudden failure. The brake caliper and master cylinder can be overhauled as described in Chapter 6.

4 Renew the hydraulic brake hoses

Yamaha recommend that the hydraulic brake hoses should be renewed every four years regardless of their apparent condition. Refer to Chapter 6 for details.

Chapter 1 Engine, clutch and gearbox

Contents

General description	1
Operations with the engine/gearbox unit in the frame	2
Operations with the engine/gearbox unit removed from the frame	3
Removing the engine/gearbox unit from the frame	4
Dismantling the engine/gearbox unit: preliminaries	5
Dismantling the engine/gearbox unit: removing the cylinder head, barrel and piston	6
Dismantling the engine/gearbox unit: removing the YPVS components	7
Dismantling the engine/gearbox unit: removing the crankcase right-hand cover	8
Dismantling the engine/gearbox unit: removing the clutch and primary drive pinion	9
Dismantling the engine/gearbox unit: removing the kickstart assembly and idler gear	10
Dismantling the engine/gearbox unit: removing the balancer shaft gears	11
Dismantling the engine/gearbox unit: removing the tachometer drive gears	12
Dismantling the engine/gearbox unit: removing the external gear selector components	13
Dismantling the engine/gearbox unit: removing the generator	14
Dismantling the engine/gearbox unit: removing the reed valve	15
Dismantling the engine/gearbox unit: separating the crankcase halves	16
Dismantling the engine/gearbox unit: removing the crankshaft and gearbox components	17
Dismantling the engine/gearbox unit: removing oil seals and bearings	18
Examination and renovation: general	19
Examination and renovation: engine cases and covers	20
Examination and renovation: bearings and oil seals	21
Examination and renovation: cylinder head	22
Examination and renovation: cylinder barrel	23
Examination and renovation: piston and rings	24
Examination and renovation: crankshaft	25
Examination and renovation: primary drive and balancer shaft gears	26
Examination and renovation: clutch	27
Examination and renovation: kickstart	28
Examination and renovation: gearbox components	29
Gearbox shafts: dismantling and reassembly	30
Reassembling the engine/gearbox unit: general	31
Reassembling the engine/gearbox unit: preparing the crankcases	32
Reassembling the engine/gearbox unit: refitting the crankshaft, balancer shaft and gearbox components	33
Reassembling the engine/gearbox unit: joining the crankcase halves	34
Reassembling the engine/gearbox unit: refitting the reed valve assembly	35
Reassembling the engine/gearbox unit: refitting the generator	36
Reassembling the engine/gearbox unit: refitting the external gear selector components	37
Reassembling the engine/gearbox unit: refitting the primary drive and balancer shaft gears	38
Reassembling the engine/gearbox unit: refitting the tachometer drive gears	39
Reassembling the engine/gearbox unit: refitting the kickstart assembly and idler gear	40
Reassembling the engine/gearbox unit: refitting the clutch	41
Reassembling the engine/gearbox unit: refitting the right-hand crankcase cover	42
Reassembling the engine/gearbox unit: refitting the YPVS components	43
Reassembling the engine/gearbox unit: refitting the piston, cylinder barrel and head	44
Refitting the engine/gearbox unit in the frame	45
Starting and running the rebuilt engine	46
Taking the rebuilt machine on the road	47

Specifications

Engine
- Capacity ..... 124 cc (7.6 cu in)
- Bore:
  - TZR and early DT models (3DB1) ..... 56.4 mm (2.22 in)
  - Later DT models (3RN1 onward) ..... 56.0 mm (2.20 in)
- Stroke:
  - TZR and early DT models (3DB1) ..... 50.0 mm (1.97 in)
  - Later DT models (3RN1 onward) ..... 50.7 mm (2.00 in)
- Compression ratio:
  - TZR models 5.9:1
  - DT models 6.8:1

Cylinder head
- Gasket face maximum warpage ..... 0.03 mm (0.0012 in)

Chapter 1 Engine, clutch and gearbox

Cylinder barrel

Standard bore ID:
- TZR and early DT models (3DB1) .... 56.40 – 56.42 mm (2.2205 – 2.2213 in)
- Later DT models (3RN1 onward) .... 56.00 – 56.02 mm (2.2047 – 2.2055 in)

Service limit:
- TZR and early DT models (3DB1) .... 56.50 mm (2.2244 in)
- Later DT models (3RN1 onward) .... 56.10 mm (2.2087 in)

Piston to bore clearance .... 0.045 – 0.050 mm (0.0018 – 0.0020 in)
Service limit .... 0.1 mm (0.004 in)
Maximum taper .... 0.05 mm (0.0020 in)
Maximum ovality .... 0.01 mm (0.0004 in)

Piston

Standard OD:
- TZR and early DT models (3DB1) .... 56.34 – 56.40 mm (2.2181 – 2.2205 in)
- Later DT models (3RN1 onward) .... 55.950 – 55.955 mm (2.2027 – 2.2029 in)

1st oversize piston:
- Later DT model (3RN4 onward) .... 56.25 mm (2.2146 in)
- All other models .... 56.65 mm (2.2303 in)

2nd oversize piston:
- Later DT model (3RN4 onward) .... 56.50 mm (2.2244 in)
- All other models .... 56.90 mm (2.2402 in)

Piston rings

	TZR and early DT models (3DB1)	Later DT model (3RN1 onward)
Top ring: Type	Keystone	Keystone
Height	1.2 mm (0.0472 in)	1.2 mm (0.0472 in)
Width	2.2 mm (0.0866 in)	2.4 mm (0.0945 in)
Second ring: Type	Plain	Plain
Height	1.2 mm (0.0472 in)	1.2 mm (0.0472 in)
Width	2.2 mm (0.0866 in)	1.9 mm (0.0748 in)

End gap – installed .... 0.30 – 0.45 mm (0.0118 – 0.0177 in)

Ring/ring groove clearance:
- Top ring .... 0.020 – 0.060 mm (0.0008 – 0.0024 in)
- Second ring .... 0.035 – 0.070 mm (0.0014 – 0.0028 in)

Crankshaft

Crank width .... 57.90 – 57.95 mm (2.2800 – 2.2815 in)
Runout limit .... 0.02 mm (0.0008 in)
Big-end axial clearance (endfloat) .... 0.20 – 0.70 mm (0.0079 – 0.0276 in)
Connecting rod small-end deflection .... 0.8 – 1.0 mm (0.0315 – 0.0394 in)

Primary drive

Reduction ratio 3.227:1 (71/22T)

Clutch

	TZR models	DT models
Number of friction plates	6	7
Number of plain plates	5	6
Number of springs	4	5

Friction plate thickness .... 2.9 – 3.1 mm (0.1142 – 0.1220 in)
Service limit .... 2.7 mm (0.1063 in)
Plain plate thickness .... 1.05 – 1.35 mm (0.0413 – 0.0531 in)
Distortion limit .... 0.05 mm (0.0020 in)
Clutch spring free length .... 34.5 mm (1.3583 in)
Service limit .... 32.0 mm (1.2598 in)
Pushrod warpage limit .... 0.15 mm (0.0059 in)

Kickstart

Spring clip tension .... 0.8 – 1.2 kg (1.76 – 2.65 lb)

Gearbox

Input and output shaft runout limit .... 0.08 mm (0.0031 in)

Ratios:

	TZR models	DT models
1st	2.833:1 (34/12T)	2.833:1 (34/12T)
2nd	1.813:1 (29/16T)	1.875:1 (30/16T)
3rd	1.368:1 (26/19T)	1.412:1 (24/17T)
4th	1.143:1 (24/21T)	1.143:1 (24/21T)
5th	1.000:1 (23/23T)	0.957:1 (22/23T)
6th	0.917:1 (22/24T)	0.818:1 (18/22T)

Final drive
Reduction ratios:
- TZR models 2.813:1 (45/16T)
- Early DT models (3DB1) 2.235:1 (55/17T)
- Later DT models (3RN1 onward) 3.563:1 (57/16T)

Chain size .... 428 ($\frac{1}{2} \times \frac{5}{16}$)

Number of links:
- TZR model 130
- DT model 134

Torque settings

Component	kgf m	lbf ft
Cylinder head nuts	2.2	16
Cylinder barrel nuts	2.8	20
Spark plug	2.0	14
Crankcase screws	0.8	5.8
Crankcase left-hand cover screws	0.5	3.6
Crankcase right-hand cover screws	0.8	5.8
Oil baffle plate screws	0.8	5.8
Crankshaft right-hand oil seal retaining plate screw	1.6	12
Input shaft right-hand bearing retaining plate screws	1.0	7.2
Tachometer driven gear housing bolt	0.5	3.6
Power valve cap screws	0.7	5.1
Power valve retaining Allen bolt	0.7	5.1
Reed valve retaining bolts	1.0	7.2
Clutch centre nut:		
TZR models	5.5	40
DT models	7.0	51
Clutch spring retaining bolts	0.6	4.3
Primary drive nut	8.0	58
Balancer shaft nut	5.5	40
Selector detent arm bolt:		
TZR models	1.0	7.2
DT models	1.4	10
Generator rotor nut	8.0	58
Generator stator screws	0.8	5.8
Transmission oil drain plug	1.5	11
Coolant drain plug	1.0	7.2
Kickstart nut	6.5	47
Gearchange lever pinch bolt:		
TZR models	1.0	7.2
DT models	1.5	11
Drive sprocket bolts – TZR models	1.0	7.2
Drive sprocket nut – DT models	6.0	43
Exhaust pipe front mounting nuts	1.8	13
Exhaust pipe rear mounting bolt – TZR models	2.5	18
Exhaust pipe mounting bolts – DT models	1.0	7.2
Engine mounting bolts – TZR models	3.2	23
Top engine bracket/frame bolt – TZR models	2.3	17
Downtube/frame bolts – TZR models	3.2	23
Front engine mounting bolt – DT models	5.8	42
Rear lower engine mounting bolt – DT models	3.3	24
Engine stay/frame bolts – DT models	3.3	24
Swinging arm pivot bolt – DT models	9.0	65

1 General description

The Yamaha TZR125 and DT125 R models employ a water cooled, single cylinder, two-stroke engine built in unit with the primary drive, clutch and gearbox. The cylinder head and barrel castings are of light alloy construction and incorporate cast-in passages for the coolant. The cylinder barrel has a cast iron liner and is fitted with the Yamaha Power Valve System, although to comply with UK legislation the power valve is pegged in place and thus cannot operate. The crankshaft is of conventional design, having needle roller bearings at the connecting rod big- and small-ends, and is supported by two ball journal main bearings. A gear driven single shaft primary balance is fitted to counter the imbalance which exists in all single-cylinder engines.

Primary drive is by helical-cut gears to the wet multi-plate clutch mounted on the end of the gearbox input shaft. The gearbox is of the six-speed constant mesh type and is lubricated by oil bath shared with the primary drive, the oil being contained in a reservoir formed by the main crankcase castings. The engine oil and water pumps are mounted on the crankcase right-hand cover and are driven from the crankshaft via the primary drive pinion and nylon idler gears.

2 Operations with the engine/gearbox unit in the frame

The following items can be removed with the engine/gearbox unit in the frame:

   (a) Cylinder head, barrel and piston
   (b) Carburettor and reed valve
   (c) Clutch assembly and primary drive gear
   (d) Oil pump
   (e) Water pump
   (f) Kickstart mechanism
   (g) Gear selector mechanism external components
   (h) Generator assembly and neutral switch
   (i) Final drive sprocket
   (j) Balancer shaft drive gears

Chapter 1 Engine, clutch and gearbox

3 Operations with the engine/gearbox unit removed from the frame

It is necessary to remove the engine/gearbox unit from the frame and to separate the crankcase halves to gain access to the following components:

(a) Crankshaft, main bearings and oil seals
(b) Gearbox clusters and bearings
(c) Gear selector drum and forks
(d) Balancer shaft

4 Removing the engine/gearbox unit from the frame

1 If the machine is dirty, wash it thoroughly before starting any major dismantling work. This will make work much easier and will rule out the risk of caked-on lumps of dirt falling into some vital component.
2 Drain the transmission oil as described in Routine maintenance. While the oil is draining remove the seat, the sidepanels, the fuel tank and the air scoops (DT models). On TZR models it will also be necessary to remove the lower fairing (if fitted).
3 Note that whenever any component is removed, all mounting nuts, bolts or screws should be refitted in their original locations with their respective washers and mounting rubbers and/or spacers.
4 Working as described in Chapter 2, drain all the coolant from the cooling system. Tilt the machine to the right to ensure that as much as possible is removed.
5 Work is made much easier if the machine is lifted to a convenient height on a purpose-built ramp or platform constructed of planks and concrete blocks. Ensure that the wheels are chocked with wooden blocks so that the machine cannot move and that it is securely tied down so that it cannot fall, also that it is supported firmly on a suitable stand.
6 Disconnect the battery (negative terminal first) to prevent any risk of short circuits. If the machine is to be out of service for some time, remove the battery and give it regular refresher charges as described in Chapter 7.
7 Remove the exhaust system. On TZR models, remove the single rear mounting bolt and the two front mounting nuts, and withdraw the complete exhaust system. On DT models, slacken the clamps on the exhaust pipe/silencer joint and remove the silencer by releasing its two mounting bolts and pulling it backwards off the exhaust main section. Remove the single mounting bolts at the front and rear, and the two front mounting nuts and manoeuvre the exhaust pipe clear of the frame. Discard the exhaust port gasket; a new gasket should be used on refitting.
8 Remove the carburettor as described in Chapter 3. Remove the oil pump cover and disconnect the oil pump control cable from its pulley and remove the cable from the crankcase cover. Then remove the oil feed pipe from the pump and swiftly plug its end with a suitable bung to stop the flow of oil. Disconnect the YEIS hose or chamber from the inlet stub and remove the chamber from the machine as described in Chapter 3.
9 Slacken the clutch cable adjuster mounted on the left-hand side of the cylinder barrel and disconnect the cable from the clutch lifting arm. Unscrew its knurled retaining ring and release the tachometer cable from its drive on the right-hand crankcase half; withdraw the cable and oil seal. Allow the clutch cable to hang down from the front of the frame and position the tachometer cable clear of the engine so neither will hinder engine removal.
10 Marking its shaft so that it can be refitted in the same position, remove the kickstart lever retaining nut and pull the lever off the shaft splines.
11 Release its clip and pull the expansion tank pipe off its union on the radiator neck. Disconnect the top and bottom radiator hoses from the engine and remove the radiator from the machine, as described in Chapter 2, so that the hoses remain attached to the radiator. Starting from the crankcase trace the generator wiring up the frame tube, releasing it from any clamps or ties securing it to the frame, then disconnect at the connectors joining it to the main wiring loom. Also disconnect the suppressor cap from the spark plug and the lead from the temperature sender.

4.7a TZR125 – remove exhaust system by releasing front mounting nuts ...

4.7b ... and remove rear mounting bolt

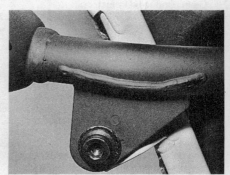
4.7c DT125 – remove mounting bolts and pull silencer free

4.7d Release two front mounting nuts ...

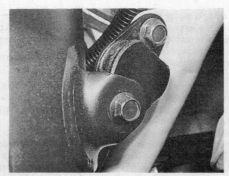
4.7e ... remove both the front ...

4.7f ... and rear mounting bolts and manoeuvre exhaust clear of frame

Chapter 1 Engine, clutch and gearbox

4.8a Disconnect YEIS hose from inlet stub ...

4.8b ... and remove chamber – TZR125 shown

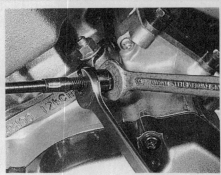
4.9 Slacken clutch cable adjuster and disconnect cable

4.10 Remove kickstart lever, marking its shaft if necessary

4.11a Disconnect generator wiring from main wiring loom – TZR125 shown

4.11b Disconnect temperature sender lead and suppressor cap

4.12 Remove gearchange pedal (see text) and left-hand crankcase cover

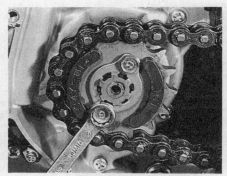
4.13a Release sprocket retaining bolts ...

4.13b ... and pull sprocket off shaft – TZR125 shown

12 Remove the gearchange pedal or linkage front arm (as appropriate) pinch bolt and pull the pedal or linkage off the gearchange shaft splines. If necessary, mark the shaft so that it can be refitted in the same position. Remove the left-hand crankcase cover retaining screws and withdraw the cover.

13 Select top gear and apply the rear brake to prevent engine rotation, then slacken both the gearbox sprocket retaining bolts (TZR models only). Remove them fully, rotate the locking plate until it can be pulled off the shaft splines and then pull off the sprocket. It may be necessary to slacken off the chain adjusters to permit this. Disengage the sprocket from the chain and allow the chain to hang over the swinging arm pivot. On DT models, the sprocket is retained by a nut and locking tab. Once the sprocket has been removed withdraw the spacer from the driveshaft oil seal.

14 The engine/gearbox unit should now be retained only by its mounting bolts. Make a final check that all components which might hinder the removal of the unit have been withdrawn.

TZR model

15 Remove the nut from the front engine mounting bolt and withdraw the bolt and spacer. Slacken and remove the five bolts which secure the right-hand downtube to the frame and lift the downtube clear of the frame.

16 Slacken the bolt which joins the cylinder head to the top engine mounting bracket and remove the bolt which secures the bracket to the frame. Do not remove the bolt which passes through the cylinder head as the engine can be hung from the top bracket prior to removal.

17 Remove both the upper and lower rear engine mounting bolts; the engine is now free to be manoeuvred out of the frame from the right-hand side.

Chapter 1 Engine, clutch and gearbox

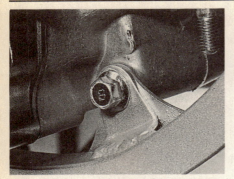
4.15a TZR125 – to release engine, remove front mounting bolt and spacer

4.15b Release all downtube retaining bolts and remove downtube

4.16 Remove both rear mounting bolts and top engine mounting/frame bolt

4.18a DT125 – remove top engine mounting bracket and front mounting bolt ...

4.18b ... followed by rear engine mounting bolt and swinging arm pivot nut ...

4.19 ... and partially withdraw pivot shaft (see text)

DT model

18 Slacken and remove the three bolts which secure the top engine mounting bracket and remove the bracket. Release both the front and lower rear mounting bolts, leaving the engine only secured by the swinging arm pivot shaft. Before removing the swinging arm pivot shaft, find a suitably sized rod or piece of tubing of the same external diameter as the pivot bolt, which can be used to support the swinging arm once the pivot bolt is partially removed.

19 Remove the nut and washer and pull the swinging arm shaft out approximately six inches, then insert the rod or piece of tubing through the frame to support the swinging arm, but not far enough to support the rear engine mounting. Slowly withdraw the swinging arm pivot shaft until the engine is seen to drop slightly. The engine is now free to be lifted out from the right-hand side of the frame. Refit the swinging arm pivot shaft so that the rear suspension is supported whilst the engine/gearbox unit is out of the frame.

5 Dismantling the engine/gearbox unit: preliminaries

1 Before any dismantling work is undertaken, the external surfaces of the unit should be thoroughly cleaned and degreased. This will prevent the contamination of the engine internals, and will also make working a lot easier and cleaner. A high flash-point solvent, such as paraffin (kerosene) can be used, or better still, a proprietary engine degreaser such as Gunk or Jizer. Use old paintbrushes and toothbrushes to work the solvent into the various recesses of the engine castings. Take care to exclude solvent or water from the electrical components and inlet and exhaust ports. The use of petrol (gasoline) as a cleaning medium should be avoided, because the vapour is explosive and can be toxic if used in a confined space.

2 When clean and dry, arrange the unit on the workbench, leaving a suitable clear area for working. Gather a selection of small containers and plastic bags so that parts can be grouped together in an easily identifiable manner. Some paper and a pen should be on hand to permit notes to be made and labels attached where necessary. A supply of clean rag is also required.

3 Before commencing work, read through the appropriate section so that some idea of the necessary procedure can be gained. When removing the various engine components it should be noted that great force is seldom required, unless specified. In many cases, a component's reluctance to be removed is indicative of an incorrect approach or removal method. If in any doubt, re-check with the text.

6 Dismantling the engine/gearbox unit: removing the cylinder head, barrel and piston

1 These items can be removed with the engine in or out of the frame but in the former case the seat, sidepanels, air scoops, fuel tank, carburettor and exhaust system must first be removed. The coolant must be drained and the suppressor cap, radiator inlet (top) hose and temperature gauge sender unit wire must be disconnected. It will also be necessary to remove the upper engine mounting bracket, which secures the cylinder head to the frame, to allow the barrel to be withdrawn. On TZR models the air filter case should also be removed as this considerably restricts access to the cylinder head and barrel. Refer to Section 4 of this chapter for further information.

2 Remove the spark plug. Working in a diagonal sequence slacken each of the cylinder head retaining nuts by $\frac{1}{4}$ of a turn until all have been loosened. Remove the nuts and washers and lift away the cylinder head. Discard the head gasket, a new one should be fitted on reassembly. If the head is stuck to the barrel do not attempt to lever it away. Break the joint by refitting the spark plug and turning the engine over, using its compression to release the head, or by tapping it gently with a soft-faced mallet. Be very careful when handling the cylinder head, it is a delicate casting and could easily be distorted or cracked by careless workmanship.

3 Bring the piston to the top of its stroke and remove the four cylinder barrel retaining nuts. Make a note of the clutch cable bracket position on the left-hand rear stud so that it can be refitted in the same position. Tap

the barrel gently with a soft-faced mallet to break the seal and lift it just enough to expose the bottom of the piston skirt. Pack a wad of clean rag in the crankcase mouth to prevent dirt or debris from falling in, then lift the barrel away. Peel the base gasket off the crankcase and unless firmly stuck in place, remove the two dowel pins for safekeeping.

4 Use a sharp-pointed instrument or a pair of needle-nosed pliers to remove one of the gudgeon pin circlips, noting exactly how it is fitted, then press out the gudgeon pin far enough to clear the connecting rod and withdraw the piston. Push out the small-end bearing. Discard the used circlips and base gasket and obtain new ones for reassembly.

5 If the gudgeon pin is a tight fit in the piston, soak a rag in boiling water, wiring it out and wrap it around the piston; the heat will expand the piston sufficiently to release its grip on the pin. If necessary, the pin may be tapped out using a hammer and drift, but take care to support firmly the piston and connecting rod while this is done.

6 The piston rings are removed by holding the piston in both hands and gently prising the ring ends apart with the thumb nails until the rings can be lifted out of their grooves and on to the piston lands, one side at a time. The rings can then be slipped off the piston and put to one side for cleaning and examination. If the rings are stuck in their grooves by excessive carbon deposits use three strips of thin metal sheet to remove them, as shown in the accompanying illustration. Be careful as the rings are brittle and will break easily if overstressed.

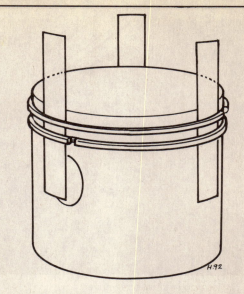

Fig. 1.1 Method of removing and refitting piston rings

Fig. 1.2 Cylinder head and barrel – TZR model (DT model similar)

1 Cylinder head
2 Nut – 5 off
3 Washer – 5 off
4 Head gasket
5 Stud – 5 off
6 Spark plug
7 Cylinder barrel
8 Base gasket
9 Stud – 2 off
10 Dowel pin – 2 off
11 Stud – 2 off
12 Clutch cable bracket
13 Nut – 4 off
14 Bolt
15 Washer
16 Bolt – 2 off
17 Left-hand valve cap
18 Gasket
19 Brass sleeve
20 Power valve
21 O-ring
22 Right-hand valve cap
23 Bolt

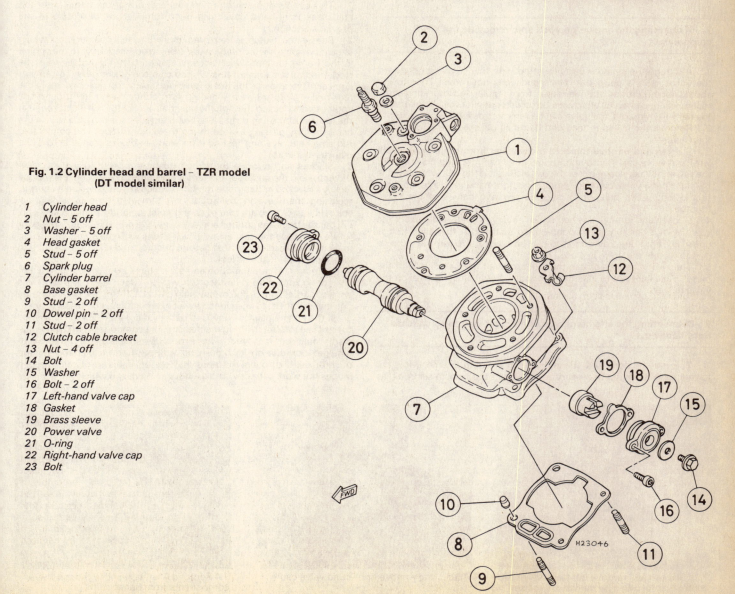

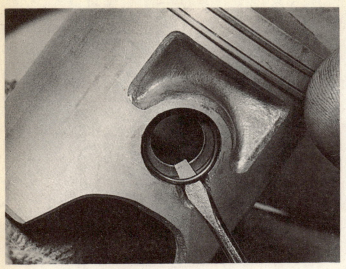

6.4 Remove circlips using a small screwdriver

described in Routine maintenance, disconnect the oil pump feed pipe and control cable, drain the coolant, remove the radiator outlet (bottom) hose and remove the kickstart lever from its shaft (see Section 4). On DT models it will also be necessary to remove the rear brake pedal with its return spring, and to remove the metal coolant pipe from the water pump casing.

2 Working in a diagonal sequence, remove all the crankcase cover screws. Store them in a cardboard template of the cover to provide a guide to their correct positions on reassembly. Also make a note of where the oil feed pipe clamp and breather hose guide are fitted so these can be refitted in their original positions.

3 Pull the cover away complete with the oil and water pumps, tapping gently with a soft-faced mallet to break the seal. Do not attempt to lever the cover away as this will almost certainly damage the sealing faces. Check that there are no thrust washers or other small components sticking to the inside of the cover. Peel off and discard the gasket and if possible remove the two locating dowels from the crankcase and store them with the casing.

7 Dismantling the engine/gearbox unit: removing the YPVS components

1 Although the cylinder barrel is fitted with the YPVS components, the valve itself is pegged in place and does not move. Therefore it should not require any attention, apart from regular decarbonising at the recommended interval, which can be carried out with the valve fitted to the barrel. If required, the valve can be removed as follows:

2 Remove the cylinder head and barrel as described in Section 6 of this Chapter.

3 Slacken and remove the bolt and washer from the centre of the left-hand valve cap. Remove both the valve cap retaining bolts and lift the cap and brass sleeve out of the barrel. Release the bolt which retains the right-hand valve cap and remove the right-hand cap to gain access to the long Allen bolt which secures the two halves of the valve together.

4 Slacken and remove this bolt whilst holding the left-hand end of the valve with a suitably sized open-ended spanner. If the Allen bolt is unusually tight, a hardwood strip can be placed through the exhaust port to wedge the valve in place.

5 The two valve halves can then be removed from each side of the barrel. Take care not to lose the two small dowel pins which align the valve halves.

8 Dismantling the engine/gearbox unit: removing the crankcase right-hand cover

1 This may be done with the engine in or out of the frame, but in the former case it will be necessary to first drain the transmission oil, as

9 Dismantling the engine/gearbox unit: removing the clutch and primary drive gear

1 This can be done with the engine in or out of the frame, after the crankcase right-hand cover has been removed as described in the previous Section.

2 Before the clutch is removed there is one consideration which should be remembered. If the work is being undertaken with the engine in the frame the nut which secures the primary drive gear must be slackened at this stage if it is wished to remove the gear. To prevent crankshaft rotation as the nut is removed, select top gear and apply the rear brake. With the rear wheel in firm contact with the ground the engine can be locked through the gear train whilst the nut is slackened. If the top end of the engine has also been removed it is possible to lock the crankshaft by passing a close fitting metal bar through the connecting rod small-end eye and resting it on wooden blocks placed across the crankcase mouth.

3 Slacken and remove the clutch pushrod locknut, using a screwdriver to hold the pushrod, and remove also the plain washer behind the locknut. Slacken and remove the bolts which secure the clutch springs, releasing them evenly by about one turn at a time in a diagonal sequence until they are free of spring pressure and can be unscrewed. Lift out the bolts and springs and withdraw the pressure plate, noting for future reference the cast arrow which aligns with the stamped circular mark in the clutch centre. Pull out the pushrod and refit the locknut and washer so that neither is lost.

4 Remove the clutch friction and plain plates from the clutch assembly noting for future reference the way in which each is fitted. This applies particularly to the metal plain plates. Lift the plates out one by one and put them to one side to await examination and reassembly.

5 If the engine/gearbox unit is to be fully dismantled, or if it is wished to examine the clutch lifting mechanism, the components of the mechanism must be removed. The simplest method is to tip the engine/gearbox unit on its right-hand side so that the ball and second pushrod will slide out of the input shaft centre. If this is not possible, or if it does not work correctly, find a long, slim, steel rod which will fit inside

7.3a Slacken and remove left-hand valve cap centre bolt followed by the valve cap itself ...

7.3b ... and remove right-hand valve cap

7.4 Remove Allen bolt and withdraw valve components from barrel

Chapter 1 Engine, clutch and gearbox

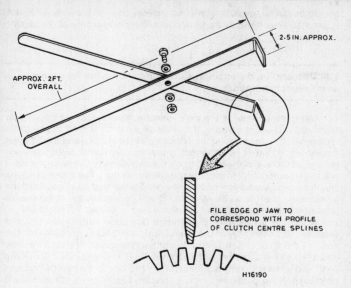

Fig. 1.3 Fabricated clutch holding tool

the input shaft. Magnetise the rod by stroking it gently across the magnets of the generator rotor and push it into the input shaft. As the rod is withdrawn gently it should bring the ball and then the pushrod with it. When the ball and pushrod have been removed, disengage the clutch lifting lever return spring from the crankcase top and lift the lever away, complete with its return spring and the plain washer.

6 Before the clutch centre nut can be removed, some method of holding the centre must be devised. Yamaha produce a special holding tool, Part Number 90890-04086, and this can be used if available. Alternatively, the home-made equivalent shown in the accompanying illustration and photo 41.3a will prove equally effective. The tool was made up from 1 in x $\frac{1}{8}$ in mild steel strip, the edges of the angled jaws being ground to fit snugly in the clutch centre splines. An assistant will be required to hold the clutch centre with the improvised tool while the nut is removed. Take care not to allow the tool to slip or the soft alloy splines will be damaged.

7 Straighten the locking tab and remove the clutch centre retaining nut and tab washer. The clutch centre can then be pulled off the input shaft, followed by the splined thrust washer and the clutch outer drum. Behind the clutch outer drum there is a further thrust washer and dished washer. Slide the thrust washer off the shaft and remove the dished washer behind the key, noting that it is fitted with its convex side towards the crankcase wall. Store all the clutch components together in a clean container to avoid losing or damaging them.

8 Slacken and remove the primary drive gear nut as described above, remove the washer and primary drive gear. Note that if the crankcases are to be split, the primary drive gear must be left in position until the balancer shaft nut has been loosened (see Section 11).

10 Dismantling the engine/gearbox unit: removing the kickstart assembly and idler gear

1 The kickstart assembly and idler gear can be removed with the engine/gearbox unit in or out of the frame, once the crankcase right-hand cover has been withdrawn as described in Section 8 of this Chapter. Note that while the kickstart shaft can be removed with the clutch assembly in situ, if the kickstart idler gear is to be removed, the clutch must be withdrawn first.

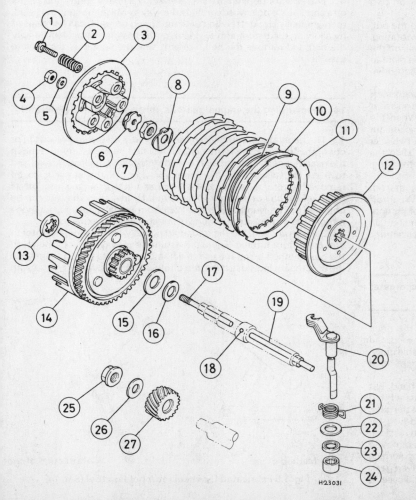

Fig. 1.4 Clutch

1 Bolt – 5 off ▲
2 Spring – 5 off ▲
3 Pressure plate
4 Locknut
5 Washer
6 Pushrod plate
7 Nut
8 Tab washer
9 Damper ring – 5 off (DT model only)
10 Plain plate – 5 off
11 Friction plate – 6 off
12 Centre
13 Splined thrust washer
14 Outer drum
15 Thrust washer
16 Dished washer
17 Left-hand pushrod
18 Steel ball
19 Right-hand pushrod
20 Clutch lifter
21 Return spring
22 Washer
23 Oil seal
24 Bearing
25 Nut
26 Washer
27 Primary drive gear
▲ 4 off on TZR models

Chapter 1 Engine, clutch and gearbox

2 To remove the kickstart assembly, use a suitable pair of pliers to unhook the outer end of the kickstart return spring from its locating hole in the crankcase wall, then allow the spring to return slowly to its relaxed position. Care is required at this point as the spring is under moderately strong tension. The kickstart assembly may then be lifted out of the crankcase.

3 Slide the white nylon spring guide down the length of the shaft and remove it. Disengage the inner end of the spring from the hole in the kickstart shaft and remove the spring, making a note of the way in which it is fitted. Slide off the large plain washer and the kickstart pinion, the latter complete with its friction clip.

4 If removal of the kickstart idler gear is required, this can be done only after the clutch has been removed. Using a pair of circlip pliers, remove the circlip from the output shaft right-hand end, then withdraw the first thrust washer, the idler gear itself, and the second thrust washer. Lastly, use the circlip pliers to remove the inner retaining circlip from the output shaft end. Because both circlips and both thrust washers are identical components, there is no need to differentiate between the inner and outer parts. Store all the kickstart components together in a clean container so that nothing is lost or damaged.

11 Dismantling the engine/gearbox unit: removing the balancer shaft gears

1 This can be done with the engine in or out of the frame. First it will be necessary to remove the right-hand crankcase cover and the clutch as described in previous Sections of this Chapter. If the top end of the engine has not been removed, a tool will be required to hold the generator rotor whilst the balancer shaft nut is slackened, see Section 14. If, however, the cylinder head, barrel and piston are removed the crankshaft can be locked by passing a close fitting bar through the small-end eye of the connecting rod and supporting the bar on two blocks of wood positioned over the crankcase mouth.

2 Straighten the balancer shaft nut tab washer and remove the nut, using one of the above methods to prevent the crankshaft from rotating. Remove the primary drive gear nut and washer and slide the gear off the crankshaft. Slacken the two screws which secure the oil baffle plate to the crankcases; this must be removed before the balancer shaft drive gear can be slid off the crankshaft.

3 Before removing the balancer shaft drive gear from the crankshaft note the stamped dot on its periphery which should align with a similar mark on the balancer shaft driven gear (see photo 38.3c) when the piston is at TDC. If these marks are indistinct they should be picked out with a blob of paint or a spirit-based felt marker before the balancer timing is disturbed. Once this point has been attended to, slide the balancer shaft drive gear off the crankshaft and prise out the long, square-section Woodruff key from the crankshaft keyway.

4 To remove the balancer shaft driven gear, lift off first the tab washer, then the gear itself. Prise the short, square-section Woodruff key from the balancer shaft keyway. Store the two groups of components, those from the crankshaft and those from the balancer shaft, in separate containers to avoid any possibility of confusion on reassembly.

12 Dismantling the engine/gearbox unit: removing the tachometer drive gears

1 The tachometer is driven from the clutch outer drum, the drive being transmitted via a white nylon drive gear to a vertically-mounted driven gear which is machined at the top to accept the drive cable inner. The whole assembly can be removed, once the crankcase right-hand cover is withdrawn, whether the clutch is in situ or not.

2 The drive gear and its shaft can be pulled out easily by hand, but note that there are two small, identical, thrust washers which are fitted one on each side of the drive gear for the purpose of location of the gear. These must not be lost or allowed to drop clear. Remove the drive gear and place the gear, thrust washers, and shaft together in a clean container. The remainder of the tachometer drive assembly is removed by unscrewing the single retaining bolt set in the crankcase top surface and by pulling away the black plastic housing and vertical driven gear. Note that if the housing is a particularly tight fit in the crankcase, it should be pulled out with a twisting motion, as if unscrewing it. Do not attempt to lever it out as the plastic moulding would only be damaged.

13 Dismantling the engine/gearbox unit: removing the external gear selector components

1 The components of the gear selector mechanism which can be removed without separating the crankcase halves consist of the gearchange shaft, incorporating the selector claw assembly, and the selector detent arm. Either component may be removed with the engine/gearbox unit in the frame or on the workbench, but in each case it will be necessary first to remove the crankcase right-hand cover and the clutch assembly, as described in previous Sections of this Chapter. If the gearchange shaft is to be removed it will be necessary, of course, to release the gearchange pedal or linkage from the shaft left-hand end.

2 When the preliminary dismantling has been carried out and the selector claw assembly of the gearchange shaft is exposed, assess the state of the shaft return spring by applying pressure first downwards, and then upwards, to the claw end of the shaft. In both cases, release the shaft and allow it to return under spring pressure to the central position before applying pressure in the opposite direction. The shaft should centre itself quickly and positively, with no trace of free play in the centre position or of weakness in the spring. If any fault is apparent the spring must be renewed on reassembly. This test is quick and easy to carry out, and must be considered essential if the engine/gearbox unit is being stripped to trace a fault in the selector mechanism.

3 The gearchange shaft can be pulled from the crankcases by hand. On DT models hold the selector fork shaft which protrudes from crankcases whilst removing the gearchange shaft. This will prevent the selector fork shaft being withdrawn along with the gearchange shaft, resulting in the selector forks falling from position.

4 The selector detent arm is tensioned by a heavy spring and pivots on a shouldered bolt. Note carefully the way in which the spring is fitted, then gradually slacken the bolt until the spring pressure can be released by allowing the detent arm to ride over the selector cam. Slacken fully the bolt and remove the bolt, the arm, and the spring as a complete assembly.

14 Dismantling the engine/gearbox unit: removing the generator

1 The generator assembly can be removed with the engine unit in or out of the frame. In the former case it will first be necessary to remove the gearchange linkage or lever and the left-hand crankcase cover, and to trace and disconnect the generator wiring if the stator is to be removed. A rotor extractor tool will be needed to draw the rotor off its taper safely; do not attempt to remove it by levering. The tool can be obtained as Part Number 90890-01189 from Yamaha dealers, but many motorcycle dealers stock a range of suitable pattern extractors.

2 Slacken and remove the rotor holding nut whilst holding the rotor to prevent it from turning. There are a number of ways to hold the rotor; the best method being the use of a home-made tool like that shown in the accompanying illustration and photo 36.3b. This was made up with

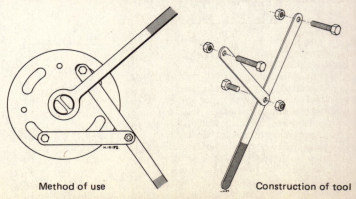

Method of use   Construction of tool

Fig. 1.5 Fabricated flywheel rotor holding tool (Sec 14)

Chapter 1 Engine, clutch and gearbox

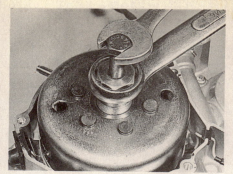

14.4 Extractor must be used to remove rotor

14.5a Neutral switch lead must be disconnected ...

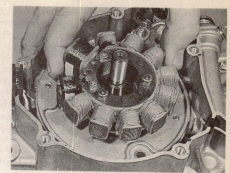

14.5b ... before stator plate is removed (TZR shown)

some steel strip and a few bolts and need not be very elaborate. If you prefer, you can buy a made-up tool of this type from Yamaha dealers as Part Number 90890-01235.
3 In the absence of a holding tool, and with the engine unit in the frame, select top gear and apply the rear brake to lock the crankshaft. If the cylinder barrel and piston have been removed, pass a smooth round bar through the connecting rod eye. Support the ends of the bar on hardwood blocks placed against the crankcase, thus locking the crankshaft.
4 With the rotor immobilised by one of the above methods, remove the nut and washer. Fit the extractor into the thread in the centre of the rotor. Holding the body of the extractor with a spanner, tighten the centre bolt to draw the rotor off its taper. If it proves stubborn, try tapping the head of the centre bolt; this will usually jar the rotor free. If necessary, tighten the bolt further and repeat the process.
5 Once the rotor has been removed, unscrew the two screws which secure the stator assembly to the crankcase and lift it away. The wiring and block connector can be freed once the wiring grommets have been displaced. Note that the neutral switch lead must also be freed. Remove the Woodruff key from the crankshaft and store it safely inside the flywheel rotor.
6 If required, the neutral switch can be unscrewed from the casing.

15 Dismantling the engine/gearbox unit: removing the reed valve

1 The reed valve can be removed with the engine in or out of the frame, but in the former case it will first be necessary to remove the seat, sidepanels, fuel tank and carburettor as described in Chapter 3.
2 Remove the four bolts which retain the inlet stub to the crankcase top surface. Lift off the inlet stub and reed valve cage. Peel off the reed valve gasket and discard it; a new gasket should be fitted on reassembly. If the engine is not to be dismantled further, plug the crankcase opening with clean rag to prevent the entry of dirt.
3 Refer to Chapter 3 for details of reed valve examination.

16 Dismantling the engine/gearbox unit: separating the crankcase halves

1 The engine/gearbox unit must be removed from the frame and all preliminary dismantling operations carried out before the crankcase halves can be separated. These operations are described in Sections 5 – 15 of this Chapter. When the engine/gearbox unit has been dismantled as described, make a final check to ensure that nothing has been omitted which might obstruct operations. Rotate the selector drum so that the protruding ears on the selector cam are aligned with the cut-outs in the crankcase wall; failure to do so will result in crankcase damage due to the selector cam contacting the casting. Place two clean wooden blocks on the workbench and support the engine/gearbox unit on them so that the crankcase right-hand side is uppermost.
2 Slacken the crankcases screws evenly and in a diagonal sequence. Before the crankcase fastening screws are removed, there is a simple device which can be used to ensure that none are lost and that each screw is refitted in its correct position. Find a sheet of cardboard and draw on it a rough outline of the crankcase right-hand side, marking the position of each of the screws. Punch holes in the cardboard at each screw position, then as each screw is removed it can be pushed into its correct corresponding hole in the cardboard and kept there until needed on reassembly.
3 Using only a soft-faced mallet, tap gently on the exposed ends of the crankshaft and gearbox shafts and all around the joint area of the two crankcase halves until initial separation is achieved. Lift off the half to be removed, ensuring it remains absolutely square so that the bearings do not stick on their respective shafts; tap gently on the shaft ends to assist removal. **Do not** use excessive force and **never** attempt to lever the cases apart. If undue difficulty is encountered, tap the cases back together and start again. If all else fails, take the assembly to an authorised Yamaha dealer for the cases to be separated using special tool Part Number 90890-01135.
4 If difficulty is encountered in achieving initial separation, it may be because corrosion has formed on the locating dowel pins. Apply a quantity of penetrating fluid to the joint area and inside the various mounting bolt passages, allow time for it to work and start again.
5 When the casing is removed, check that there are no loose components such as dowels or thrust washers which might drop clear and be lost. Any such components should be refitted in their correct locations. Unless the two locating dowels are firmly fixed in the right-hand crankcase half they should be removed and stored in a safe place.

17 Dismantling the engine/gearbox unit: removing the crankshaft and gearbox components

1 When the crankcase halves have been separated, the crankshaft, the balancer shaft, and the gearbox components may be removed.

16.1 Rotate selector drum to enable it to pass through cutouts in crankcase

2 Start by pulling out the front and rear selector fork shaft. Pivot all three selector forks about their respective pinions so that the guide pins come away from the selector drum and provide sufficient clearance for the drum to be withdrawn. This is done by pulling the drum out of its locating hole in the crankcase.
3 Withdraw the selector forks, taking great care to note exactly where each one is fitted and in what way it is fitted. It is useful to degrease each fork as it is withdrawn and to mark it with a spirit-based felt pen as a guide to correct assembly, but note that the selector forks fitted to the machine featured in the photographs were identified by numbers and model codes cast in each side, and if such is found to be the case on the machine being worked on, these marks will be sufficient provided notes are made of their position. As each fork is identified and withdrawn, replace it on its shaft to prevent any confusion on reassembly.
4 The gear clusters should be removed now. Carefully pull them out of their bearings, treating both as a single assembly. It may be necessary to tap on the output shaft left-hand end to free the shaft from its bearing.
5 The balancer shaft can be lifted out easily by hand, leaving only the crankshaft to be removed from the crankcase left-hand side. The manufacturer recommends the use of the same special tool used for separating the crankcase halves. The two mounting bolts of the tool should be threaded into the drilled drilled and tapped holes of the two bosses cast on either side of the left-hand main bearing housing. The tool centre bolt is then tightened down on the crankshaft left-hand end and is used to press the crankshaft out. If, as is likely, the tool is not available, an alternative method is given which requires only a soft-faced mallet and some care.
6 To use this method, refit the rotor nut on the crankshaft end to avoid damaging the thread and place the crankcase on two wooden blocks situated as close around the crankshaft as possible to give maximum support. The blocks should be of a size to hold the crankcase half far enough from the workbench to allow the crankshaft to be removed. Using a soft-faced mallet with one hand, and supporting the crankshaft with the other, carefully tap the crankshaft out of its housing. Do not use excessive force and do not allow the crankshaft to drop away.
7 Put all the removed components to one side to await examination and reassembly.

18 Dismantling the engine/gearbox unit: removing oil seals and bearings

1 Before removing any oil seal or bearing, check that it is not secured by a retaining plate. If this is the case, use an impact driver or spanner, as appropriate, to release the securing screws or bolts and lift away the retaining plate.
2 Oil seals are easily damaged when disturbed and, thus, should be renewed as a matter of course during overhaul. Prise them out of position using the flat of a screwdriver whilst taking care not to damage the alloy seal housings; note which way round the seals are fitted.
3 It is possible to remove the right-hand main bearing oil seal without separating the crankcases; either by screwing in two self-tapping screws so that the seal can be pulled out with two pairs of pliers, or by simply digging the seal out with a sharply-pointed instrument. This method is **not recommended** as it is difficult to carry out without scratching the seal housing or crankshaft or without damaging the main bearing. Furthermore, it is almost impossible to fit a new seal without damaging it and nothing can be done to trace and rectify the fault which caused the seal to fail in the first place.
4 The crankshaft and gearbox bearings are a press fit in their respective crankcase locations. To remove a bearing, the crankcase casting must be heated so that it expands and releases its grip on the bearing, which can then be drifted or pulled out.
5 To prevent casting distortion, it must be heated evenly to a temperature of about 100°C by placing it in an oven; if an oven is not available, place the casting in a suitable container and carefully pour boiling water over it until it is submerged. Note that if heat is used in this way, care must be taken first to remove any components from the casting which would be destroyed by excess heat, e.g. the neutral switch.
6 Taking care to prevent personal injury when handling heated components, lay the casting on a clean surface and tap out the bearing using a hammer and a suitable drift. If the bearing is to be re-used apply the drift only to the bearing outer race, where this is accessible, to avoid damaging the bearing. In some cases it will be necessary to apply

18.2 Oil seals should be removed with a flat-bladed screwdriver

pressure to the bearing inner race; in such cases closely inspect the bearing for signs of damage before using it again. When drifting a bearing from its housing it must be kept square to the housing to prevent tying in the housing with the resulting risk of damage. Where possible, use a tubular drift such as a socket spanner which bears only on the bearing outer race; if this is not possible, tap evenly around the outer race to achieve the same result.
7 In cases where bearings are pressed into blind holes, they may be removed by heating the casting and tapping it face downwards on to a clean wooden surface to dislodge the bearing under its own weight. If this is not successful the casting should be taken to a motorcycle service engineer who has the correct internally expanding bearing puller. If a bearing sticks to its shaft on removal, it can only be safely removed using a knife-edged bearing puller.

19 Examination and renovation: general

1 Before any component is examined, it must be cleaned thoroughly. Being careful not to mark or damage the item in question, use a blunt-edged scraper (an old kitchen knife or a broken plastic ruler can be very useful) to remove any caked-on deposits of dirt or oil, followed by a good scrub with a soft wire brush (a brass wire brush of the type sold for cleaning suede shoes is best, with an assortment of bottle-cleaning brushes for ports, coolant passages etc). Take care not to remove any paint code marks from internal components.
2 Soak the component in a solvent to remove the bulk of the remaining dirt or oil. If one of the proprietary engine degreasers (such as Gunk or Jizer) is not available, a high flash-point solvent such as paraffin (kerosene) should be used. The use of petrol as a cleaning agent cannot be recommended because of the fire risk. With all of the above cleaning agents take great care to prevent any drops getting into the eyes and try to avoid, prolonged skin contact. To finish off the cleaning procedure wash each component in hot soapy water (as hot as your hands can bear); this will remove a surprising amount of dirt on its own and the residual heat usually dries the component very effectively. Carefully scrape away any remaining traces of old gasket material from all joint faces.
3 Check all coolant passages and oilways for blockages, using compressed air to clear them, or implements such as pipe cleaners.
4 If there is the slightest doubt about the lubrication system, for example if a fault appears to have been caused by a failure of the oil supply, all components should be dismantled so that the oilways can be checked and cleared of any possible obstructions. Refer to Chapter 3 for details of the lubrication system. Always use clean, lint-free rag for cleaning and drying components to prevent the risk of small particles obstructing oilways.
5 Examine carefully each part to determine the extent of wear, checking with the tolerance figures listed in the Specifications section of

Chapter 1 Engine, clutch and gearbox

this Chapter. If there is any doubt about the condition of a particular component, play safe and renew.

6 Various instruments for measuring wear are required, including an internal and external micrometer or vernier gauge, and a set of standard feeler gauges. Additionally, although not absolutely necessary, a dial gauge and mounting bracket are invaluable for accurate measurement of endfloat, and play between components of very low diameter bores – where a micrometer cannot reach. After some experience has been gained, the state of wear of many components can be determined visually, or by feel, and a decision on their suitability for re-use can be made without resorting to direct measurement.

20 Examination and renovation: engine cases and covers

1 Small cracks or holes in aluminium castings may be repaired with an epoxy resin adhesive, such as Araldite, as a temporary measure. Permanent repairs can only be effected by argon-arc welding, and only a specialist in this process is in a position to advise on the economics or practicability of such a repair.

2 Damaged threads can be economically reclaimed by using a diamond section wire insert, of the Helicoil type, which is easily fitted after drilling and re-tapping the affected thread. Most motorcycle dealers and small engineering firms offer a service of this kind.

3 Sheared studs or screws can usually be removed with screw extractors, which consist of tapered, left-hand thread screws, of very hard steel. These are inserted by screwing anticlockwise into a pre-drilled hole in the stud, and usually succeed in dislodging the most stubborn stud or screw. If a problem arises which seems to be beyond your scope, it is worth consulting a professional engineering firm before condemning an otherwise sound casing. Many of these firms advertise regularly in the motorcycle papers.

4 If gasket or other mating surfaces are marked or damaged in any way they can be reclaimed by rubbing them on a sheet of fine abrasive paper laid on an absolutely flat surface such as a sheet of plate glass. Start with 200 grade paper and finish with 400 grade and oil. Use a gentle figure-of-eight pattern, maintaining light but even pressure on the casting. Note that if large amounts of material are to be removed, advice should be sought as to the viability of re-using the casting in question; the internal clearances are minimal in many cases between rotating or moving components and the castings. Stop work as soon as the entire mating surface is polished by the action of the paper.

5 Note that the mating surface may become distorted outwards around the mounting screw holes, usually because these have been grossly overtightened. In this event, use a large drill bit or countersink to very lightly skim the raised lip from around the screw hole, then clean up the whole surface as described above.

6 Finally, check that all screw or bolt tapped holes are clean down to the bottom of each hole; serious damage can be caused by forcing a screw or bolt down a dirty thread and against an incorrect stop caused by the presence of dirt, oil, swarf or blobs of old jointing compound. At the very least the component concerned will be incorrectly fastened, at worst the casting could be cracked. The simplest way of cleaning such holes is to use a length of welding rod or similar to check that the hole is clean all the way to the bottom and to dig out any embedded foreign matter, then to give each hole a squirt of contact cleaner or similar solvent applied from an aerosol via the long plastic nozzle usually supplied. Be careful to wear suitable eye protection while doing this; the amount of dirt and debris that can be ejected from each hole is surprising.

21 Examination and renovation: bearings and oil seals

1 Ball bearings should be washed thoroughly to remove all traces of oil then tested as follows. Hold the outer race firmly and attempt to move the inner race up and down, then from side to side. Examine the bearing balls, cages and tracks, looking for signs of pitting or other damage. Finally spin the bearings hard; any roughness caused by wear or damage will be felt and heard immediately. If any free play, roughness or other damage is found the bearing must be removed.

2 Roller bearings are checked in much the same way, except that free play can be checked only in the up and down direction with the components concerned temporarily reassembled. Remember that if a roller bearing fails it may well mean having to replace, as well as the bearing itself, one or two components which form its inner and outer races. If in doubt about a roller bearing's condition, renew it.

3 Do not waste time checking oil seals; discard all seals and O-rings disturbed during dismantling work and fit new ones on reassembly. Considering their habit of leaking once disturbed, and the amount of time and trouble necessary to replace them, they are relatively cheap if renewed as a matter of course.

22 Examination and renovation: cylinder head

1 Remove all traces of carbon from the cylinder head, using a blunt-ended scraper. Finish by polishing with metal polish, to give a smooth, shiny surface. This will aid gas flow and will also prevent carbon from adhering so firmly in the future.

2 Check the condition of the threads in the spark plug hole. If the threads are worn or stretched as the result of overtightening the plug, they can be reclaimed by a 'Helicoil' thread insert. Most dealers have the means of providing this cheap but effective repair.

3 Inspect the water passages cast into the cylinder head, and where necessary remove any accumulated corrosion or scale. This can result from failure to use the recommended coolant mixture. Be sure to remove any debris from the passages by flushing them through with clean water.

4 Leakage between the head and barrel will indicate distortion. Check the head by placing a straightedge across several places on its mating surface and attempting to insert a 0.03 mm (0.0012 in) feeler gauge between the two. Remove excessive distortion by rubbing the head mating surface in a slow circular motion against emery paper placed on plate glass (see Section 20).

5 Note that most cases of cylinder head distortion can be traced to unequal tensioning of the cylinder head securing nuts or to tightening them in the incorrect sequence.

23 Examination and renovation: cylinder barrel

1 The usual indication of a badly worn cylinder barrel and piston is piston slap, a metallic rattle that occurs when there is little or no load on the engine.

2 Carefully remove the ring of carbon from the bore mouth so that bore wear can be accurately assessed and check the barrel/cylinder head mating surface as described in the previous Section. Clean all carbon from the exhaust port and all traces of old gasket from the cylinder base.

3 Examine the bore for scoring or other damage, particularly if broken rings are found. Damage will necessitate reboring and a new piston regardless of the amount of wear. A satisfactory seal cannot be obtained if the bore is not perfectly finished.

4 There will probably be a lip at the uppermost end of the cylinder bore which marks the limit of travel of the top piston ring. The depth of the lip will give some indication of the amount of bore wear that has taken place even though the amount of wear is not evenly distributed.

5 The most accurate method of measuring bore wear is by the use of a cylinder bore DTI (Dial Test Indicator) or a bore micrometer. Measure at the top (just below the wear ridge), middle and bottom of the bore, in line with the gudgeon pin axis and at 90° to it, avoiding port areas. Take six measurements in all noting each one carefully. Take the maximum figure obtained to indicate bore wear; if this is equivalent to, or greater than the service limit given in the Specifications Section of this Chapter, the barrel must be rebored and the next oversize piston and rings must be obtained. This can be confirmed by subtracting the overall diameter of the piston from the maximum cylinder bore figure obtained. If the difference calculated exceeds 0.1 mm (0.004 in) the bore can be regarded as excessively worn, assuming that the piston is unworn (see Section 24).

6 Alternatively, insert the piston into the bore in its correct position with the skirt just below the wear ridge, i.e. at the point of maximum wear. Measure the gap using feeler gauges. If the piston/cylinder clearance is 0.1 mm (0.004 in) or more the barrel must be rebored. It

must be stressed that this can only be a guide to the degree of bore wear alone unless the piston is known to be unworn; have the barrel measured accurately by an authorised Yamaha dealer or similar expert before a rebore is undertaken.

7   After a rebore has been carried out, the reborer should hone the bore lightly to provide a fine cross-hatched surface so that the new piston and rings can bed in correctly. Also the edges of the ports should be chamfered first with a scraper, then with fine emery, to prevent the rings from catching on them and breaking.

8   If a new piston and/or rings are to be run in a part-worn bore the surface should be prepared first by glaze-busting. This involves the use of a special honing attachment with (usually) an electric drill to provide a surface similar to that described above. Most motorcycle dealers have such equipment and will be able to carry out the necessary work for a small charge.

7   Measure ring wear by inserting each ring into part of the bore which is unworn and measuring the gap between the ring ends with a feeler gauge. If the measurement exceeds that given in Specifications, renew the ring. Use the piston crown to locate the ring squarely in the bore approximately 20 mm (0.8 in) from the gasket surface.

8   Reject any rings which show discoloured patches on their mating surfaces; these should be brightly polished from firm contact with the cylinder bore.

9   Do not assume when fitting new rings that their end gaps will be correct; the installed end gap must be measured as described above to ensure that it is within the specified tolerances; if the gap is too wide another piston ring set must be obtained (having checked again that the bore is within specified wear limits), but if the gap is too narrow it must be widened by the careful use of a fine file.

24 Examination and renovation: piston and rings

1   Disregard the existing piston and rings if a rebore is necessary; they will be replaced with oversize items. Note also that it is considered a worthwhile expense by many mechanics to renew the piston rings as a matter of course, regardless of their apparent condition.

2   Measure the piston diameter at right angles to the gudgeon pin axis, at a point 10 mm (0.4 in) above the base of the skirt. If the piston is worn to or beyond the service limit given in the Specifications Section of this Chapter it must be renewed. Note that oversize pistons and rings are available in two sizes. The amount of oversize is stamped on the piston crown; do not forget to include it in any calculations.

3   Note that the manufacturer's alternative method of determining whether a rebore is necessary is to subtract the piston overall diameter from the maximum bore internal diameter measurement obtained. Since it is possible for the piston to have worn at a greater rate than the bore, this can only be accurate if the piston measurement is that of an unworn component (ie new). If this method is used, always have your findings checked by an expert before undertaking a rebore; it may suffice to renew the piston alone.

4   Piston wear usually occurs at the skirt, especially on the forward face, and takes the form of vertical score marks. Reject any piston which is badly scored or which has been badly blackened as the result of the blow-by of gas. Slight scoring of the piston can be removed by careful use of a fine swiss file. Use chalk to prevent clogging of the file teeth and the subsequent risk of scoring. If the ring locating pegs are loose or worn, renew the piston.

5   The gudgeon pin should be a firm press fit in the piston. Check for scoring on the bearing surfaces of each part and where damage or wear is found, renew the part affected. The circlip retaining grooves must be undamaged; renew the piston rather than risk damage to the bore through a circlip becoming detached. Discard the circlips themselves; these should **never** be reused.

6   Any built-up of carbon in the ring grooves can be removed by using a section of broken piston ring, the end of which has been ground to a chisel edge. Using a feeler gauge, measure each ring to groove clearance. Renew the piston if the measurement obtained exceeds that given in Specifications and if the rings themselves appear unworn.

25 Examination and renovation: crankshaft

1   Big-end failure is characterised by a pronounced knock which will be most noticeable when the engine is worked hard. The usual causes of failure are normal wear, or failure of the lubrication supply. In the case of the latter, the noise will become apparent very suddenly, and will rapidly worsen.

2   Check for wear with the crankshaft set in the TDC (top dead centre) position, by pushing and pulling the connecting rod. No discernible movement will be evident in an unworn bearing, but care must be taken not to confuse endfloat, which is normal, and bearing wear.

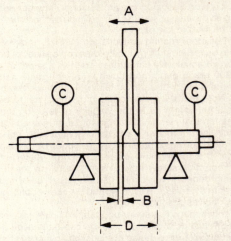

Fig. 1.6 Crankshaft measurement points

A   Big-end bearing radial clearance
B   Big-end bearing axial clearance
C   Crankshaft runout
D   Crankshaft width across flywheels

24.2 Measuring piston outside diameter

24.7 Measuring piston ring end gap

25.3 Measuring crankshaft big-end endfloat

Chapter 1 Engine, clutch and gearbox

the bearing if its roller cage is cracked or worn.
5   It should be noted that crankshaft repair work is of a highly specialised nature and requires the use of equipment and skills not likely to be available to the average private owner. Such work should not be attempted by anyone without this equipment and the skill to use it.

26 Examination and renovation: primary drive and balancer shaft gears

1   The primary drive consists of a crankshaft pinion which engages a large gear mounted on the inner face of the clutch drum. Both components are relatively lightly loaded and will not normally wear until a very high mileage has been covered.
2   If wear or damage is discovered it will be necessary to renew the component concerned. In the case of the large driven gear it will be necessary to purchase a complete clutch drum because the two items form an integral unit and cannot be obtained separately. Check for play in the outer drum/driven gear by holding the clutch outer drum and attempting to twist or rotate the primary driven gear backwards and forwards. Unfortunately no figures are given with which to assess the amount of freeplay, and it will therefore be a matter of experience to decide whether renewal is necessary or not. If in any doubt, take the clutch outer drum to an authorised Yamaha dealer for inspection.
3   Check that there is no obvious damage to the balancer shaft drive gears such as chipped or broken teeth, if any wear is found the gears should be renewed as a pair.

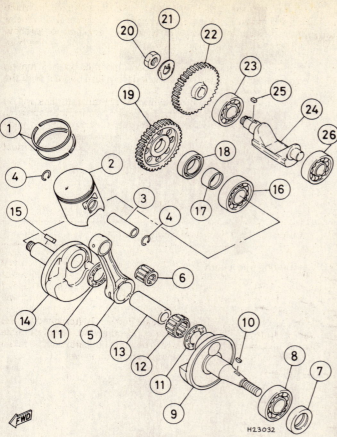

Fig. 1.7 Crankshaft and balancer shaft

1	Piston rings	14	Right-hand crankshaft half
2	Piston	15	Woodruff key
3	Gudgeon pin	16	Right-hand main bearing
4	Circlip – 2 off	17	Spacer
5	Connecting rod	18	Oil seal
6	Small-end bearing	19	Balancer shaft drive gear
7	Oil seal	20	Nut
8	Left-hand main bearing	21	Tab washer
9	Left-hand crankshaft half	22	Balancer shaft driven gear
10	Woodruff key	23	Bearing
11	Thrust bearing – 2 off	24	Balancer shaft
12	Big-end bearing	25	Woodruff key
13	Crankpin	26	Bearing

27 Examination and renovation: clutch

1   After an extended period of service, the friction plates will have become worn sufficiently to warrant renewal, to avoid subsequent problems with clutch slip. The lining thickness is measured across the friction plate using a vernier caliper. If any plate is worn to or beyond the specified service limit, the friction plates must be renewed as a complete set. Note that if new friction plates are fitted, they must be coated with a light film of the recommended transmission oil.
2   The plain plates should be free from any signs of blueing, which would indicate that the clutch had overheated in the past. Check each plate for distortion by laying it on a flat surface, such as a sheet of plate glass or similar, and measuring any detectable gap using feeler gauges. The plates must be less than the specified limit out of true, and should be renewed as a complete set if any are found to be distorted beyond the set limit.
3   The clutch springs may, after a considerable mileage, require renewal, and their free length should be checked as a precautionary measure. If any spring is found to less than the specified limit in length the clutch springs must be renewed as a set.
4   On DT models examine the clutch damper ring for any sign of wear or damage and renew it if necessary.
5   Check the condition of the slots in the outer surface of the clutch centre and the inner surfaces of the outer drum. In an extreme case, clutch chatter may have caused the tongues of the friction plates to make indentations in the slots of the outer drum, or the tongues of the plain plates to indent the slots of the clutch centre. These indentations will trap the clutch plates as they are freed and impair clutch action. If the damage is only slight the indentations can be removed by careful work with a file and the burrs removed from the tongues of the clutch plates in similar fashion. More extensive damage will necessitate renewal of the parts concerned.
6   Examine the pushrods for wear and distortion, rolling them on a sheet of plate glass or other flat surface and using feeler gauges to measure any discernible gap. If bent, the pushrods may be straightened, but if incorrect adjustment or lack of lubrication have caused such a build-up of friction that the hardening on the pushrod ends has broken down, the pushrods and the steel ball must be renewed. Such a case will be easy to see due to the heavy blueing and distorted shape which will result. The steel ball plays an important role in that it permits the

3   If measuring facilities are available, set the crankshaft in V-blocks and check the big-end radial clearance using a dial gauge mounted on a suitable stand. Big-end radial clearance is measured as the amount of lateral deflection at the connecting rod small-end eye, thus magnifying any wear present in the bearing. Refer to the accompanying illustration if clarification of this point is required. Crankshaft runout can also be checked with the V-blocks and dial gauge; the gauge pointer being applied to the straight, non-tapered, part of each mainshaft. Big-end axial clearance (endfloat) is measured with feeler gauges of the correct thickness which should be a firm sliding fit between the thrust washer next to the big-end eye and the machined shoulder on the flywheel. The crank width should also be checked. This is done by measuring the distance between the two outer edges of the crank webs. If any clearance exceeds the service limits given in the specifications Section of this Chapter the crankshaft assembly should be taken to an authorised Yamaha dealer or similar repair agent for repair or renewal.
4   Push the small-end bearing into the connecting rod eye and push the gudgeon pin through the bearing. Hold the rod steady and feel for movement between it and the pin. If movement is felt, renew the pin, bearing or connecting rod as necessary so no movement exists. Renew

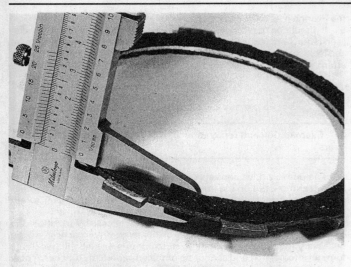

27.1 Measuring clutch friction plate thickness ...

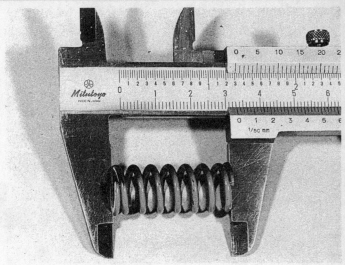

27.3 ... and spring free length

27.5a Examine slots in clutch centre ...

27.5b ... and outer drum for burrs and indentations

right-hand pushrod to rotate with the clutch assembly, but does not transmit this rotary motion to the left-hand pushrod, thus protecting the lifting cam of the lever assembly. Check that the ball is perfectly round. Any flaws in the ball will mean that it must be renewed. All these clutch lifting components must be well greased on reassembly.

28 Examination and renovation: kickstart

1 The kickstart mechanism is a robust assembly and should not normally require attention. Apart from obvious defects such as a broken return spring, the friction clip is the only component likely to cause problems if it becomes worn or weakened. The clip is intended to apply a known amount of drag on the kickstart pinion, causing the latter to run up its quick thread and into engagement when the kickstarter lever is operated.
2 The clip can be checked using a spring balance. Hook one end of the balance onto the looped end of the friction clip. Pull on the free end of the balance and note the reading at the point where pressure overcomes the clip's resistance. If the reading is outside the specified limits and the mechanism has been malfunctioning, renew the clip as a precaution. Do not attempt to adjust a worn clip by bending it.
3 Examine the kickstart pinion for wear or damage, remembering to check it in conjunction with the output shaft-mounted idler pinion. In view of the fact that these components are not subject to continuous use, a significant amount of wear or damage is unlikely to be found.

29 Examination and renovation: gearbox components

1 Give the gearbox components a close visual inspection for signs of wear or damage such as broken or chipped teeth, worn dogs, damaged or worn splines and bent selectors. Renew any parts found unserviceable because they cannot be reclaimed in a satisfactory manner.
2 The gearbox shafts are unlikely to sustain damage unless the lubricating oil has been run low or the engine has seized and placed an unusually high loading on the gearbox. Check the surfaces of the shaft, especially where a pinion turns on it, and renew the shaft if it is scored or has picked up.
3 The gearbox shafts can be checked for trueness by removing all the components from the shaft, as described in Section 30, and setting the shaft up in V-blocks. Slowly turn the shaft whilst measuring any runout with a dial gauge. If the runout exceeds the specified limit the shaft must be renewed.

Chapter 1 Engine, clutch and gearbox

29.5 Examine selector forks and ...

29.6 ... drum for signs of wear

4 Check the selector fork shafts for straightness by rolling them on a sheet of plate glass. A bent shaft will cause difficulty in selecting gears and will make the gearchange action particularly heavy. Similarly, check the gearchange shaft for straightness.
5 The selector forks should be examined closely, to ensure that they are not bent or badly worn. Wear is unlikely to occur unless the gearbox has been run for a period with a particularly low oil content.
6 The tracks in the selector drum, with which the selector forks engage, should not show any undue signs of wear unless neglect has led to under lubrication of the gearbox. Check the bearing fitted to the right-hand end of the selector drum for any sign of wear. This bearing cannot be removed from the selector drum and the complete selector drum assembly must be renewed if the bearing is faulty.
7 Examine the gear selector claw assembly noting that worn or rounded ends on the claw can lead to imprecise gear selection. Similarly check the change pins in the selector drum end. The springs in the selector mechanism and the detent or stopper arm should be unbroken and not distorted or bent in any way.

30 Gearbox shafts: dismantling and reassembly

1 The gearbox clusters should not be disturbed needlessly, and need only be stripped when careful examination of the whole assembly fails to reveal the cause of a problem, or where obvious damage, such as stripped or chipped teeth is discovered.
2 The input and output shaft components should be kept separate to avoid confusion during reassembly. Using circlip pliers, remove the circlip and plain washer which retain each part. As each item is removed, place it in order on a clean surface so that the reassembly sequence is obvious and the risk of parts being fitted the wrong way round or in the wrong sequence is avoided. Care should be exercised when removing circlips to avoid straining or bending them excessively. The clips must be opened just sufficiently to allow them to be slid off the shaft. Note that Yamaha recommend that all the circlips should be renewed regardless of their condition. Examine all thrust washers for any signs of wear or damage, renewing them if necessary.
3 Having examined the gearbox components as described in the previous Section, reassemble each shaft as described below, referring to the accompanying line drawing and photographs for guidance. Note that when fitting a circlip to a splined shaft, the circlip ears must be positioned in the middle of one of the splines to ensure that the circlip is as secure as possible on its shaft. Ensure that the bearing surfaces of each component are liberally oiled before fitting.
4 If problems arise in identifying the various gear pinions which cannot be resolved by reference to the accompanying photographs and illustrations, the number of teeth on each pinion will identify them. Count the number of teeth on the pinion and compare this figure with that given in the Specifications Section of this Chapter, remembering that the output shaft pinions are listed first, followed by those on the input shaft. The problem of identification of the various components should not arise, however, if the instructions given in paragraph 2 of this Section are followed carefully.

Input shaft
5 The input shaft is easily identified by its integral 1st gear pinion. Hold the threaded end of the shaft (right-hand end) and slide the components on from the opposite end (left-hand end) using the following procedure.
6 Slide the 5th gear pinion along the shaft, with its dogs on the left-hand side of the pinion, followed by a thrust washer. Secure the gear and thrust washer with a circlip. Slide the 3rd/4th gear pinion onto

30.6a Take the bare input shaft and slide on the 5th gear pinion ...

30.6b ... followed by a thrust washer and circlip

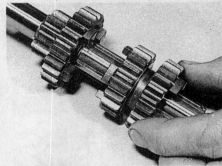

30.6c Refit 3rd/4th gear pinion ...

Chapter 1 Engine, clutch and gearbox

30.6d ... secure it with a circlip and fit a thrust washer

30.6e Slide on the 6th gear pinion ...

30.6f ... followed by the 2nd gear pinion and secure them with a circlip

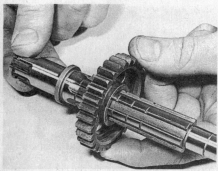
30.8a Take the bare output shaft and slide on the 2nd gear pinion ...

30.8b ... followed by a thrust washer and circlip

30.8c Refit 6th gear pinion and secure it with another circlip

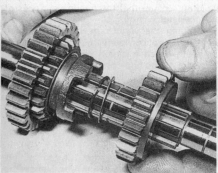
30.8d Refit a thrust washer followed by the 4th gear pinion ...

30.8e ... and secure them both with a circlip

30.8f Slide on the 3rd gear pinion and thrust washer ...

30.8g ... and secure them with a circlip

30.8h Refit 5th gear pinion ...

30.8i ... followed by the 1st gear pinion

Chapter 1 Engine, clutch and gearbox

30.8j Secure with a thrust washer and circlip

the shaft so that the slightly smaller of the two gears is on the right-hand side of the pinion. Fit a circlip to the second groove in from the left-hand end of the shaft followed by a thrust washer and the 6th gear pinion (dogs on the right-hand side of the pinion). Finally refit the 2nd gear pinion and secure it with a circlip.

Output shaft

7 Hold the splined end of the shaft (left-hand end) and slide the components on from the opposite end (right-hand end) using the following procedure.

8 Slide the 2nd gear pinion onto the shaft, making sure that its recessed face is on the right-hand side. Refit the thrust washer and secure the thrust washer and gear with a circlip. Slide the 6th gear pinion onto the shaft, so that the gear itself is positioned next to the 2nd gear pinion, and secure with a circlip. Refit the splined thrust washer and 4th gear pinion (flat surface on the right-hand side), securing them with another circlip, followed by the 3rd gear pinion (flat surface on the left-hand side), splined thrust washer and circlip. Slide the 5th gear pinion onto the shaft, so that the selector fork groove faces the 3rd gear pinion, and refit the 1st gear pinion. Note the oilways in the raised centre area on one side of the 1st gear pinion; these should be on the right-hand side of the gear. Finally refit the thrust washer securing it with a circlip.

31 Reassembling the engine/gearbox unit: general

1 Before reassembly of the engine/gearbox unit is commenced, the various component parts should be cleaned thoroughly and placed on a sheet of clean paper, close to the working area.
2 Make sure all traces of old gaskets have been removed and that the mating surfaces are clean and undamaged. Great care should be taken when removing old gasket compound not to damage the mating surface. Most gasket compounds can be softened using a suitable solvent such as methylated spirits, acetone or cellulose thinner. The type of solvent required will depend on the type of compound used.

Fig. 1.8 Gearbox shafts

1 Input shaft right-hand bearing
2 Bearing retainer
3 Screw – 2 off
4 Input shaft
5 Input shaft 5th gear
6 Thrust washer – 2 off
7 Circlip – 3 off
8 Input shaft 3rd/4th gear
9 Input shaft 6th gear
10 Input shaft 2nd gear
11 Input shaft left-hand bearing
12 Output shaft right-hand bearing
13 Circlip
14 Thrust washer
15 Output shaft 1st gear
16 Output shaft 5th gear
17 Circlip – 4 off
18 Splined thrust washer 2 off
19 Output shaft 3rd gear
20 Output shaft 4th gear
21 Output shaft 6th gear
22 Thrust washer
23 Output shaft 2nd gear
24 Output shaft
25 Output shaft left-hand bearing
26 Oil seal
27 Gearbox sprocket ▲
28 Locking plate ▲
29 Bolt – 2 off ▲
30 Spacer △
31 Gearbox sprocket △
32 Tab washer △
33 Nut △
▲ TZR models
△ DT models

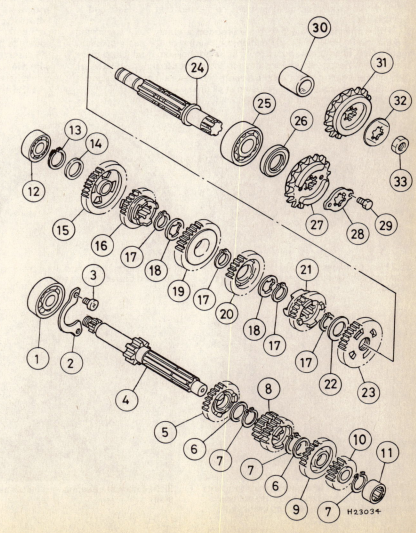

Gasket compound of the non-hardening type can be removed using a soft brass-wire brush of the type used for cleaning suede shoes. A considerable amount of scrubbing can take place without fear of harming the mating surfaces. Some difficulty may be encountered when attempting to remove gaskets of the self-vulcanising type, the use of which is becoming widespread, particularly as cylinder head and base gaskets. The gasket should be pared from the mating surface using a scalpel or a small chisel with a finely honed edge. Do not, however, resort to scraping with a sharp instrument unless necessary.

3   Gather together all the necessary tools and have available an oil can filled with clean engine oil. Make sure that all new gaskets and oil seals are to hand, also all new parts required. Nothing is more frustrating than having to stop in the middle of a reassembly sequence because a vital gasket or replacement has been overlooked. As a general rule each moving engine component should be lubricated thoroughly as it is fitted into position.

4   Make sure that the reassembly area is clean and that there is adequate working space. Refer to the torque and clearance settings wherever they are given. Many of the smaller bolts are easily sheared if overtightened. Always use the correct size screwdriver bit for the cross-head screws and never an ordinary screwdriver or punch. If the existing screws show evidence of maltreatment in the past, it is advisable to renew them as a complete set.

32   Reassembling the engine/gearbox unit: preparing the crankcases

1   At this stage the crankcase castings should be clean and dry with any damage, such as worn threads, repaired. If any bearings are to be refitted, the crankcase casting must be heated first as described in Section 18.

2   Place the heated casting on a wooden surface, fully supported around the bearing housing. Position the bearing on the casting, ensuring that it is absolutely square to its housing then tap it fully into place using a hammer and a tubular drift such as a socket spanner which bears only on the bearing outer race. Be careful to ensure that the bearing is kept absolutely square to its housing at all times.

3   Oil seals are fitted into a cold casing in a similar manner. Apply a thin smear of grease to the seal circumference to aid the task, then tap the seal into its housing using a hammer and a tubular drift which bears only on the hard outer edge of the seal, thus avoiding any risk of the seal being distorted. Tap each seal into place until its flat outer surface is just flush with the surrounding crankcase, or against its locating shoulder, where appropriate. Oil seals are fitted with the spring-loaded lip towards the liquid (or gas) being retained, ie with the manufacturer's marks or numbers facing outwards. Note, make sure that the right-hand main bearing double-lipped) seal is fitted correctly. If the seal is examined closely, it will be seen that on one side the seal is marked OUTSIDE. The seal should be fitted with this side facing outwards, away from the crankshaft.

4   Where retaining plates are employed to secure bearings or oil seals such as the input shaft right-hand bearing retainer and the crankshaft right-hand oil seal retainer, thoroughly degrease the threads of the mounting screws, apply a few drops of thread-locking compound to them, and tighten them to the specified torque settings.

5   When all bearings and oil seals have been fitted and secured, lightly lubricate the bearings with clean engine oil and apply a thin smear of grease to the sealing lips of each seal.

6   Support the crankcase left-hand half on two wooden blocks placed on the work surface; there must be sufficient clearance to permit the crankshaft and gearbox components to be fitted.

33   Reassembling the engine/gearbox unit: refitting the crankshaft, balancer shaft and gearbox components

1   Refit the primary drive nut to protect the crankshaft threaded end and insert the crankshaft as far as possible into its main bearing in the left-hand crankcase half, using a smear of oil to ease the task. Align the connecting rod with the crankcase mouth, check that the crankshaft is square to the crankcase and support the crank webs at a point opposite the crankpin to prevent distortion while the crankshaft is driven home with a few firm blows from a soft-faced mallet. Do not risk damaging the crankshaft by using excessive force; if undue difficulty is encountered, take the assembly to an authorised Yamaha dealer for the crankshaft to

32.2 Heat crankcase casting before fitting bearings, taking care to prevent scalding the hands

32.3 Ensure right-hand main bearing oil seal is fitted correctly (see text)

32.4 Apply thread-locking compound to all seal and bearing retaining plate screws

33.1 Fit crankshaft and gearbox components into left-hand crankcase half – do not use excessive force when refitting crankshaft

33.3a Lubricate bearing and install balancer shaft

33.3b Fit gearbox shaft assemblies as a single unit

Chapter 1 Engine, clutch and gearbox

33.4a Fit selector forks using marks or notes made on dismantling

33.4b Swing forks away to provide clearance for selector drum to be installed

33.4c Refit selector forks into their respective slots in the drum and install selector fork shafts

be drawn into place using the correct service tool.
2   When the crankshaft is fitted, remove the protecting nut and check that the crankshaft revolves easily with no trace of distortion.
3   Check that the balancer shaft left-hand bearing is lubricated and insert the balancer shaft into it. Check that the shaft revolves easily and freely. Check that the gearbox input and output shaft left-hand bearings are lubricated and that the oil seal next to the output shaft bearing has plenty of grease on its sealing lips. Temporarily wrap a layer of tape over the output shaft splines to protect the oil seal lips on installation, then mesh both shafts together, ensuring that all matching gear pinions are correctly mated, and insert the shafts into their bearings treating the two as a single assembly. Take great care not to damage the oil seal as the output shaft splines pass through it, and tap gently on the shafts'

right-hand ends, if necessary, to drive them into place.
4   Refit the selector forks to their respective pinion grooves, using the notes made on dismantling for guidance. If none were made note that on the machine shown in the photographs marks were provided in the form of a number cast into one surface of each fork. The fork marked 1 is fitted to the output shaft 6th gear pinion (bottom fork), that marked 3 is fitted to the output shaft 5th gear pinion (top fork) and the fork marked 2 is fitted to the input shaft 3rd/4th gear pinion. All forks are fitted with the numbers facing towards the left-hand crankcase (downwards). Install the selector drum and refit the selector forks into their respective grooves in the drum. Refit the selector fork shafts.
5   When all components are fitted, check that the selector drum is in the neutral position and that both shafts are free to rotate, then rotate the drum to check that all six gears can be selected with relative ease. Return to the neutral position and lubricate thoroughly all bearings and bearing surfaces.

34 Reassembling the engine/gearbox: joining the crankcase halves

1   Apply a thin film of sealing compound to the gasket surface of the left-hand crankcase half, then press the two locating dowels firmly into their recesses in the crankcase mating surface. Make a final check that all components are in position and that all bearings and bearing surfaces are lubricated.
2   Lower the upper crankcase half into position, using firm hand pressure only to push it home. It may be necessary to give a few gentle taps with a soft-faced mallet to drive the casing fully into place. Do not use excessive force; instead check that all shafts and dowels are correctly fitted and accurately aligned, and that the crankcase halves are exactly square to each other. If necessary, pull away the upper crankcase half to rectify the problem before starting again.
3   When the two halves have joined correctly and without strain, refit the crankcase retaining screws, using the cardboard template to position each screw correctly. Working in a diagonal sequence from the centre outwards, progressively tighten the screws until all are securely and evenly fastened to the specified torque setting.
4   Wipe away any excess sealing compound from around the joint area, check the free running and operation of the crankshaft and gearbox components. If a particular shaft is stiff to rotate, a smart tap on each end using a soft-faced mallet will centralise the shaft in its bearing. If this does not work, or if any other problem is encountered, the crankcases must be separated again to find and rectify the fault. Pack clean rag into the crankcase mouth to prevent the entry of dirt, then refit the transmission oil drain plug, if removed, tightening it to the specified torque setting.

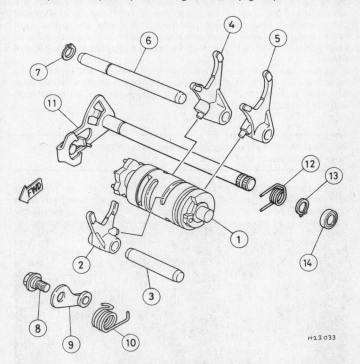

Fig. 1.9 Gear selector mechanism

1   Selector drum
2   Selector fork – input shaft 3rd/4th gear
3   Selector fork shaft
4   Selector fork – output shaft 5th gear
5   Selector fork – output shaft 6th gear
6   Selector fork shaft
7   Circlip
8   Pivot bolt
9   Detent arm
10 Return spring
11 Gearchange shaft (DT model shown)
12 Return spring
13 Circlip
14 Oil seal

35 Reassembling the engine/gearbox unit: refitting the reed valve assembly

1   Refit the reed valve and inlet stub assembly to the crankcases using a new gasket. Tighten the four mounting bolts in a diagonal sequence,

60   Chapter 1 Engine, clutch and gearbox

34.1 Ensure crankcase locating dowels are in position (arrowed) and refit right-hand crankcase half

35.1a Using a new gasket fit the reed valve assembly ...

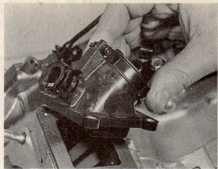

35.1b ... and inlet stub – tighten mounting bolts securely

to the recommended torque setting.

2 If the work is being carried out with the engine/gearbox unit in the frame, it will now be necessary to reassemble all those components dismantled as described in Section 15 of this Chapter, the necessary work being described in Sections 45 and 46 of this Chapter.

36 Reassembling the engine/gearbox unit: refitting the generator

1 Degrease the rotor and crankshaft tapers and refit the Woodruff key into the crankshaft.

2 Relocate the stator in its original position, using the wiring as a guide, and press the rubber wire sealing grommets into their cutouts in the crankcase wall. Refit and tighten the two stator mounting screws to the specified torque setting, having applied a few drops of thread-locking compound to their threads. If previously removed, refit the neutral switch and its sealing washer in the casing. Reconnect the neutral switch wire.

3 Degrease the rotor and crankshaft mating surfaces and remove any metal particles or swarf from the rotor magnets. Align the slot in the rotor taper with the Woodruff key fitted to the crankshaft and gently push the rotor onto the crankshaft. Gently tap the rotor centre with a soft-faced hammer to seat it and fit the plain washer and nut. Lock the crankshaft using one of the methods described in Section 14 and tighten the nut to the specified torque setting.

37 Reassembling the engine/gearbox unit: refitting the external gear selector components

1 Assemble the selector detent arm and its return spring together

using the accompanying photographs and the notes made on dismantling. Apply a few drops of thread-locking compound to the threads of the shouldered pivot bolt and fit all three parts as a single assembly. Screw the bolt in as far as possible, then pull or lever the detent arm so that the roller on its tip comes into correct contact with the selector cam. Check that the spring is correctly positioned, then tighten the pivot bolt to its specified torque setting. Check that the detent arm is free to move easily.

2 If the gearchange shaft return spring was removed for any reason it must now be refitted. Ensure that the spring is fitted the correct way round and that its ends engage correctly with the protruding tang of the selector claw plate. Refit the retaining circlip using a suitable pair of pliers.

3 Wrap the splines at the end of the gearchange shaft in a strip of PVC insulating tape or use a heavy application of grease to protect the sealing lips of the oil seal in the crankcase left-hand half. Carefully insert the gearchange shaft assembly and push it into place, taking great care to use a slight twisting action as the shaft splines pass through the oil seal. This will minimise the risk of damaging the delicate sealing lips of the seal.

4 When the shaft is in place, move it upwards and downwards, releasing it each time, to check the operation of the selector mechanism. The shaft should return quickly and positively to the centre position with no trace of free play or weakness.

38 Reassembling the engine/gearbox unit: refitting the primary drive and balancer shaft gears

1 When refitting the primary drive and balancer shaft gear pinions it will be necessary to lock the crankshaft to permit the retaining nuts to be tightened. This is best achieved by placing a close-fitting round bar

36.1 Fit Woodruff key to the crankshaft ...

36.2 ... and refit stator plate assembly. Apply thread-locking compound to the retaining screws

Chapter 1 Engine, clutch and gearbox

36.3a Fit rotor followed by plain washer and nut ...

36.3b ... and lock crankshaft as shown whilst nut is tightened to specified torque setting

through the connecting rod small-end eye and resting the bar on two wooden blocks placed across the crankcase mouth, although if the work is being carried out with the engine/gearbox unit in the frame, the method used on dismantling will be sufficient.

2   Grease the right-hand main bearing oil seal lip and carefully refit the spacer. Insert the long square-section Woodruff key into the crankshaft keyway and refit the balancer shaft drive gear to the crankshaft, making sure that the timing punch mark is on the outward face. Refit the primary drive gear, washer and nut. Tighten the nut to the specified torque setting.

3   Insert the short, square-section Woodruff key into the balancer shaft keyway and turn the crankshaft to the TDC position so that the timing mark on the balancer shaft drive gear faces the balancer shaft itself. Partially locate the balancer shaft driven gear over the Woodruff key, making sure that the timing mark dot is on the outward face. Rotate the driven gear until its timing mark dot aligns with the corresponding dot on the drive gear, then push the drive gear fully into position so that the two sets of teeth mesh and check that the two timing marks are correctly aligned.

4   When the balancer shaft timing is correctly set, refit a new tab washer and the nut to the balancer shaft. Note that the tab washer must be installed so that the tang on the rear of the washer fits into the keyway of the balancer shaft. Lock the crankshaft and tighten the balancer shaft nut to the specified torque setting. Using a hammer and a

37.1 Apply thread-locking compound to shouldered bolt and refit detent arm and return spring

37.2 Ensure gearchange shaft return spring is correctly fitted

37.4 Check the operation of the external gear selector components as described in text

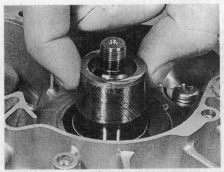
38.2a Grease right-hand main bearing oil seal and fit spacer

38.2b Long square section Woodruff key fits in crankshaft keyway

38.2c Fit balancer shaft drive gear making sure punch mark faces outwards ...

38.2d ... followed by primary drive gear ...

38.2e ... washer and nut

38.2f Lock the crankshaft and tighten nut to specified torque setting

38.3a Fit short square-section Woodruff key to balancer shaft keyway

38.3b Locate balancer shaft driven gear on key ...

38.3c ... and rotate gear until both punch marks align

38.4a Fit a new tab washer ...

38.4b ... and tighten nut to specified torque setting – secure with the tab washer

38.5 Fit oil baffle plate having first applied thread-locking compound to its retaining screws

suitable punch, lock the driven gear retaining nut by bending a portion of its tab washer up against one of the flats of the nut.

5 Finally refit the oil baffle plate to the crankcases having first applied a few drops of thread-locking compound to the threads of its retaining screws. Tighten the screws to the specified torque setting.

39 Reassembling the engine/gearbox unit: refitting the tachometer drive gears

1 When refitting the tachometer drive assembly, lightly oil the vertical driven gear and insert it down through the opening in the crankcase top surface into its housing, then check the condition of the sealing O-ring fitted around the spigot of the black plastic housing. Renew the O-ring if necessary, apply a thin smear of grease to the housing spigot and fit the housing with a twisting motion as if screwing it in; use grease to assist the task. When the housing is fully in place, fit and tighten the single retaining bolt to the specified torque setting. Do not overtighten the bolt, or the housing will crack. If the clutch is not in place the white nylon drive gear may be assembled on its shaft with a thrust washer on each side and offered up as a complete assembly.

2 If, however, the clutch is in place, a slightly different procedure must be used. Place one of the thrust washers on the crankcase wall around the housing machined to accept the drive gear shaft, and use a smear of grease to stick it in place. Insert the drive gear, ensuring that it meshes correctly with both the clutch outer drum and the vertical driven gear, and hold it in place while pushing the shaft into place. The second thrust washer may then be fitted over the shaft end to rest against the drive gear. Take great care to check that the assembly is still in place before the crankcase right-hand cover is refitted, as it is only the latter which retains the drive gear assembly.

Chapter 1 Engine, clutch and gearbox

39.1a Fit driven gear and plastic housing. Avoid overtightening the retaining bolt

39.1b Fit a plain thrust washer ...

39.1c ... followed by plastic drive pinion and another plain thrust washer

40 Reassembling the engine/gearbox unit: refitting the kickstart assembly and idler gear

1 Using a suitable pair of circlip pliers, refit the circlip on the innermost of the two grooves machined in the output shaft right-hand end. Place a thrust washer over the shaft end, followed by the kickstart idler gear; note that one side of the idler gear has a protruding boss while the other side is heavily chamfered around its periphery. It is the chamfered side which must face outwards. Secure the idler gear with the second thrust washer and a circlip.

2 If the kickstart assembly was dismantled for repair work, it must now be rebuilt and offered up to the crankcase as a single unit. Place the kickstart pinion on the workbench with its flat surface downwards and install the friction clip so that its looped end points upwards. Place the kickstart pinion over the end of the kickstart shaft with the friction clip pointing inwards, towards the shaft left-hand end, and slide the pinion down the shaft to engage with the large diameter splines, then refit the large plain washer which locates against the shaft shoulder. Place the kickstart return spring over the shaft with both spring ends facing inwards, towards the large diameter splines, and insert the spring short inner end into its locating hole in the kickstart shaft. Refit the white nylon spring guide onto the shaft, ensuring that the cut-out on one end is facing inwards and engages with the spring inner end.

3 Lubricate all bearing surfaces of the kickstart assembly and insert the assembly into the crankcase. As the end of the shaft fits into its housing in the crankcase wall, ensure that the large shaft stopper arm locates correctly against the stop cast in the crankcase wall, and that the friction clip looped end fits into the recess provided for it. Make a quick check to ensure that the kickstart assembly is operating correctly by rotating the shaft as far as possible anticlockwise. The shaft should rotate smoothly and easily until the stopper arm comes into contact with its second stop, and the kickstart pinion should move easily up its shaft splines to engage with the teeth of the idler gear. Return the shaft to its normal position by rotating it clockwise and use a suitable pair of pliers to bring the return spring long hooked end around so that it can be

40.1a Fit a circlip to output shaft right-hand end ...

40.1b ... followed by first plain thrust washer and idler gear

40.1c Fit second plain thrust washer and secure with another circlip

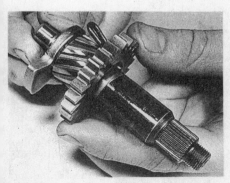
40.2a Note direction of kickstart friction clip

40.2b Slide large plain washer into position

40.2c Fit return spring ensuring short inner end locates in hole in kickstart shaft

40.2d Slide nylon guide into place ensuring that notch engages with inner end of kickstart spring

40.3a Install kickstart assembly ensuring friction clip is correctly located in recess provided in casing ...

40.3b ... then engage the kickstart return spring into its locating hole

engaged in its locating hole in the crankcase wall. Take care at this point to ensure that the spring end is hooked firmly into the locating hole, as the spring will be under quite heavy pressure.

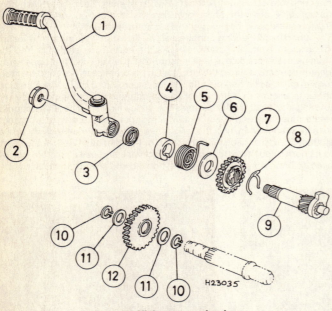

Fig. 1.10 Kickstart mechanism

1	Kickstart lever	7	Kickstart pinion
2	Nut	8	Friction clip
3	Oil seal	9	Kickstart shaft
4	Spring guide	10	Circlip – 2 off
5	Return spring	11	Thrust washer – 2 off
6	Washer	12	Idler gear

41 Reassembling the engine/gearbox unit: refitting the clutch

1 Commence reassembly by fitting the dished washer and thrust washer over the input shaft right-hand end, noting that the dished washer is fitted with its convex side towards the crankcase wall.

2 Lubricate the clutch outer drum centre bush, the input shaft and the thrust washer. Place the clutch outer drum over the input shaft end and slide it into place, ensuring that the kickstart idler gear and primary drive gear mesh correctly with their respective sets of teeth on the outer drum. Refit the second, splined, thrust washer with its slightly chamfered surface facing outwards, then fit the clutch centre over the shaft end. Fit a new tab washer with its small protruding tang located in the cut-out in the clutch centre boss, whereupon the retaining nut may be replaced on the shaft end and tightened down as far as possible by hand.

3 Lock the clutch centre using the method employed on dismantling and tighten the retaining nut to the recommended torque setting. Secure the nut by bending a portion of the tab washer up against one of the flats of the nut. Check that the clutch outer drum and the clutch centre can rotate freely, easily, and independently of one another. Thoroughly grease the left-hand pushrod and insert it into the input shaft, followed by the steel ball.

4 The clutch plain and friction plates should be coated with transmission oil prior to installation. It will be noted that each of the plain plates has a part of its outer edge machined off. This effectively makes the plate become slightly out of balance, causing each plate to be thrown outwards under the centrifugal force and thus reducing clutch noise. To prevent the whole clutch from getting out of balance it is necessary to arrange the plates so that the machined areas are spaced evenly around its circumference. This can be achieved by arranging each cutaway area to be approximately 60° (DT models) and 72° (TZR models) away from the previous one. (See Fig. 1.11.) Starting with a friction plate, and then a plain plate, insert the plates in succession until all are fitted. Note on DT models that the friction plate with the larger internal diameter, should be the second friction plate to be installed into the clutch outer drum, along with the clutch damper ring.

41.1a Fit dished washer to input shaft right-hand end ...

41.1b ... followed by a thrust washer

41.2a Refit clutch outer drum ...

Chapter 1 Engine, clutch and gearbox

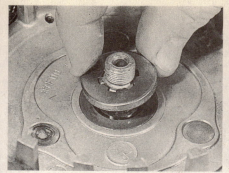
41.2b ... followed by the splined thrust washer ...

41.2c ... and clutch centre

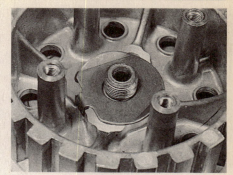
41.2d Fit a new tab washer ...

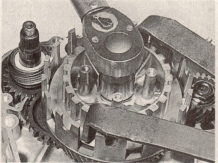
41.3a ... and tighten nut to specified torque setting whilst holding clutch centre as shown

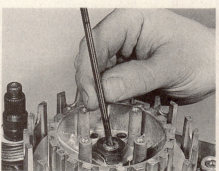
41.3b Grease clutch left-hand pushrod and insert it into input shaft ...

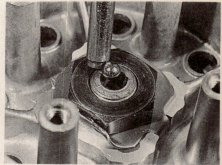
41.3c ... followed by the steel ball

5 When fitting the clutch pressure plate, note that the arrow mark cast on it must align with the stamped circular mark on the periphery of the clutch centre. Assemble the right-hand pushrod, pushrod plate, washer and locknut on the clutch pressure plate; secure the locknut by hand only. Thoroughly grease the pushrod, insert it into the input shaft, and guide the pressure plate into place, ensuring that it fits over the clutch spring pillars and aligns with the clutch centre mark at the same time. Refit the clutch springs and their retaining bolts. Secure the assembly by tightening the bolts evenly and in a diagonal sequence to the specified torque setting.

6 The clutch lifting mechanism must now be correctly set up, as described in Routine maintenance.

42 Reassembling the engine/gearbox unit: refitting the right-hand crankcase cover

1 Make a final check that all components are correctly in place, securely fastened, and lubricated. Ensure that the two locating dowels are securely in place in the gasket face of the crankcase right-hand half and check that the tachometer drive pinion outer thrust washer is still in position.

2 Thoroughly clean both gasket surfaces and place a new gasket on the crankcase sealing face, using a smear of grease to stick it in place. Smear a small amount of grease on the sealing lips of the kickstart shaft oil seal, and liberally grease the splines of the kickstart shaft itself. Offer up the cover, taking great care not to damage the oil seal lips as the kickstart shaft passes through them. Ensure also that the water pump and oil pump drive pinions engage correctly with the crankshaft pinions, rotating the crankshaft if necessary to achieve this.

3 Once the cover is fully home, refit the retaining screws in their correct locations, not forgetting the oil feed pipe clamp and breather hose guide. Tighten the retaining screws to the specified torque setting in progressive stages and working from the centre outwards in a diagonal sequence.

4 If the work is being carried out with the engine/gearbox unit in the frame, it will now be necessary to reassemble all those components dismantled as described in Section 8 of this Chapter, the necessary work being described in Sections 45 and 46 of this Chapter.

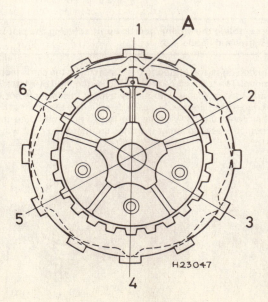

Fig. 1.11 Correct arrangement of clutch plain plate notches – DT model (Sec 41) Note also punch mark stamped on clutch centre – A

Chapter 1 Engine, clutch and gearbox

41.4 Note cutaway area on plain plates. These must be arranged as described in text

41.5a Align arrow mark on pressure plate with circular mark on clutch centre and fit pressure plate

41.5b Refit springs and retaining bolts and tighten them to the specified torque setting

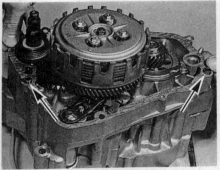
42.1 Ensure both dowels are in position (arrowed) and fit a new gasket

42.2 Rotate crankshaft whilst refitting cover to ensure oil and water pump drive pinions engage correctly

42.3 Do not omit oil pipe clamp or breather hose guide when refitting retaining screws

43 Reassembling the engine/gearbox unit: refitting the YPVS components

1 Check that the power valve components are clean and free from carbon deposits. Refit the two dowel pins to one half of the valve and offer up the valve halves as shown in the accompanying photograph. Join the two halves together, making sure that both dowel pins locate correctly. Refit the long Allen bolt which secures the valve, having first applied a few drops of thread-locking compound to its threads. Hold the flats on the left-hand end of the valve with a suitably sized open-ended spanner and tighten the Allen bolt to the specified torque setting.

2 Refit the brass sleeve to the left-hand valve cap so that the peg on the inside of the valve cap is seated in the groove of the brass sleeve. Refit the cap and sleeve assembly in the cylinder barrel using a new gasket, and refit the two cap retaining Allen screws. Note that it may be necessary to rotate the valve to correctly align it with the brass sleeve before the valve cap will seat properly. Note, before proceeding any further check that the valve is correctly positioned (see accompanying photographs) as it is possible to lock the valve 180° out of position which will drastically affect performance. If the valve is found to be positioned incorrectly simply remove the valve cap and rotate the valve 180° before refitting it. Refit the valve cap centre bolt and the right-hand valve cap and Allen screw. Tighten all screws to their recommended torque settings.

44 Reassembling the engine/gearbox unit: refitting the piston, cylinder barrel and head

1 The piston rings are refitted to the piston using the same technique as on removal. The top surface of each ring is identified by a letter which is either stamped or etched into it. However, if this letter is not visible the top surface of the ring can be easily identified by the fact that the end gap is chamfered and is wider at the top to accommodate the ring locating peg. It is essential that the rings are fitted the right way up in

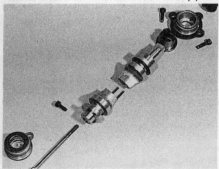
43.1a YPVS components

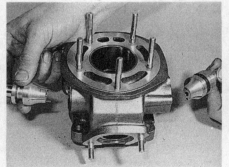
43.1b Refit the valve as shown taking care not to drop dowel pins

43.2a Ensure slot on brass sleeve locates with peg on inside of left-hand end cap

Chapter 1 Engine, clutch and gearbox

43.2b Refit sleeve and end cap assembly ...

43.2c ... and check valve is correctly positioned ...

43.2d ... and is not 180 degrees out of position

their respective grooves and that their end gaps are located correctly by the retaining pegs. The top ring is of the Keystone type and is of tapered section and the second ring is of plain section.

2 Lubricate the crankshaft main bearings and big-end bearing with two-stroke oil, then lubricate the small-end bearing and insert this into the connecting-rod small-end eye. Check that the sealing surfaces of the crankcase mouth and the cylinder barrel are completely clean and free from grease, oil, or dirt, then lightly grease a new cylinder base gasket. Check that the two locating dowels are in place around the cylinder right-hand mounting studs and lower the gasket into place over the studs and dowels, ensuring that it aligns correctly with the ports. Pack the crankcase mouth with a piece of clean rag.

3 Position the piston over the connecting rod small-end eye so that the arrow cast in the piston crown points to the front, or towards the exhaust port, and secure it by pushing the gudgeon pin through the small-end bearing and the piston itself. Fit new circlips ensuring that they are correctly seated. *Do not allow the open ends of the circlip to be positioned in the slot provided for circlip removal.*

4 Check that the piston rings are still correctly positioned in their grooves and then lubricate both the piston rings and cylinder bore with fresh two-stroke oil. Lower the cylinder barrel over the piston whilst pressing the piston ring ends together. This should not require excessive force so if any difficulty is encountered remove the barrel and double check that the rings are correctly fitted before trying again. Once both rings are fully engaged in the bore, remove the rag from the crankcase mouth and push the barrel carefully down to rest on the

44.2a Lubricate all bearings thoroughly – note rag packing crankcase mouth

44.2b Ensure dowels are in position (arrowed) and fit a new gasket

44.3 Fit piston ensuring that arrow on piston crown faces forwards

44.4a Ensure piston ring end gaps align with piston locating pegs as shown

44.4b Do not forget clutch cable bracket when refitting barrel retaining nuts

44.5a Fit a new head gasket ...

44.5b ... and tighten cylinder head nuts to specified torque settings

crankcase mouth. Refit the clutch cable bracket on the cylinder left-hand rear mounting stud, then fit and tighten the four cylinder retaining nuts. The nuts should be tightened in a diagonal sequence to their recommended torque setting.

5 Rotate the crankshaft so that the piston is at the top of its stroke and wipe away any surplus oil. Check that the mating surfaces of the cylinder head and barrel are clean and dry, then place a new cylinder head gasket in position over the studs. Refit the cylinder head, plain washers and domed nuts. Tighten the cylinder head nuts evenly in a diagonal sequence, working up in stages until the specified torque setting is reached.

6 If the work is being carried out with the engine/gearbox unit in the frame, it will now be necessary to reassemble all those components dismantled as described in Section 6 of this Chapter, the necessary work being described in Sections 45 and 46 of this Chapter.

45 Refitting the engine/gearbox unit in the frame

1 The engine/gearbox is refitted to the frame by reversing the removal sequence. Position the machine securely, preferably upright, rather than on its side stand, and ease the engine into place. Align the engine with its mountings and refit the mounting bolts from the left-hand side. On DT models grease the swinging arm pivot shaft with a high melting-point lithium-based grease before sliding it through the rear engine mounting boss. On TZR models, once the engine is correctly aligned, refit the right-hand frame tube. Do not forget to fit the spacer to the left-hand side of the engine on the front mounting bolt. On all models make sure all mounting bolts and washers are correctly positioned before tightening their nuts (where applicable) to their specified torque settings.

2 On DT models, grease the lips of the output shaft oil seal and carefully refit the spacer. Engage the gearbox sprocket on the chain and fit the sprocket over the output shaft splines. Refit the tab washer and tighten the nut to the specified torque setting, whilst locking the engine as described in Section 4. Secure the nut by bending the locking tab up against one of the flats on the nut. On TZR models, refit the sprocket and slide the locking plate along the shaft, rotating it so that its bolt holes align with those of the sprocket and its internal splines are locked in the groove cut in the shaft splines. Apply thread-locking compound to their threads and refit the two retaining bolts; lock the engine to prevent rotation while they are tightened to the specified torque setting. On all models refit the final drive chain and adjust the chain tension, rear brake adjustment and stop lamp switch setting as described in Routine maintenance. If the chains joining link was disconnected, the spring clip's closed end must face the normal direction of chain travel on refitting.

45.2 Refit gearbox sprocket and final drive chain - DT model shown

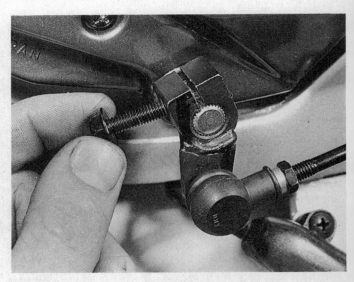
45.3 Refit gearchange pedal using the marks made on dismantling to position it correctly

45.7a Connect clutch cable to lifting arm and bend up tab (arrowed)

45.7b Do not omit oil seal when refitting tachometer drive cable

3 Refit the left-hand crankcase cover, using a new gasket and tighten the screws in a diagonal sequence. Using the marks made on dismantling, refit the gearchange pedal or linkage front arm (as appropriate) to the gearchange shaft splines and tighten the pinch bolt.
4 Route the main generator lead from the crankcase up to the frame, connect it again to the main loom and secure it with any clamps or ties provided. Refit the spark plug having first set its gap as described in Routine maintenance. Reconnect the suppressor cap and temperature sender unit lead to the cylinder head. Referring to Chapter 2, refit the radiator to the machine and connect the top and bottom hoses to their respective stubs on the engine. Reconnect the expansion tank pipe to the union on the radiator neck.
5 Refit the kickstart lever on the shaft splines using the marks made on dismantling to position it correctly. Check that the kickstart does not make contact with the crankcase cover and refit the lever retaining nut, tightening it to the specified torque setting.
6 Refit the carburettor as described in Chapter 3 and connect the oil pump control cable to the pump. Working as described in Routine maintenance, check the throttle cable and the pump cable adjustment. Connect the oil tank/pump feed pipe – again ensuring that it is correctly routed – and bleed any air from the feed pipe and pump using the bleed screw, as described in Chapter 3. Refit the YEIS chamber to the machine and connect the hose or chamber (as appropriate) to the inlet stub.
7 Connect the clutch cable to the clutch lifting arm and remount the cable on the bracket situated on the left-hand rear cylinder barrel nut. Adjust the cable as described in Routine maintenance. Insert the tachometer cable into its housing on the top of the right-hand crankcase, not forgetting to first refit the oil seal. It might prove necessary to turn the engine over on the kickstart to engage the drive and cable. Once the cable is correctly engaged with its drive, securely tighten its knurled retaining ring.
8 Refit the exhaust system. Use grease to stick a new gasket to the exhaust port and fit the mountings loosely. Tighten the exhaust port fasteners first, then the remaining mountings. On DT125 models, do not forget to check the tubular seal between the exhaust pipe and silencer; this must be renewed if damaged, split or broken. Tighten all exhaust mounting bolts and nuts to the specified torque settings.
9 Connect the battery to the main loom (negative terminal last), checking its terminals and the vent hose as described in Routine maintenance. Check that the transmission oil drain plug is fastened to the specified torque setting and pour in 800 cc (1.4 pint) of SAE 10W/30 SE engine oil; if this amount cannot be measured accurately use the sightglass as described in Routine maintenance but note that the level will drop after the engine has been run for the first time as the oil is distributed around the various components and will therefore require careful re-checking. Refit the filler plug.
10 Refill the radiator and expansion tank as described in Chapter 2 but do not yet refit the radiator cap as the coolant level will fall when the engine is run for the first time.
11 Refit the fuel tank as described in Chapter 3, turn on the fuel tap and check for any leaks which must be cured immediately.
12 Make a final check that all components have been refitted and that all are correctly adjusted and securely fastened; the exceptions are the components of the engine lubrication and cooling systems which are still to receive a final check before the machine is ready for use.

46 Starting and running the rebuilt engine

1 Attempt to start the engine using the usual procedure adopted for a cold engine. Do not be disillusioned if there is no sign of life initially. A certain amount of perseverance may prove necessary to coax the engine into activity even if new parts have not been fitted. Should the engine persist in not starting, check that the spark plug has not become fouled by the oil used during re-assembly. Failing this go through the fault diagnosis section and work out what the problem is methodically.
2 When the engine does start, keep it running as slowly as possible to allow the oil to circulate. Open the choke as soon as the engine will run without it. During the initial running, a certain amount of smoke may be in evidence due to the oil used in the reassembly sequence being burnt away. The resulting smoke should gradually subside.
3 Check the engine for blowing gaskets and coolant or oil leaks. Before using the machine on the road, check that all the gears select properly, and that the controls function correctly. As soon as the engine has warmed up to normal operating temperature and is running smoothly stop it and check the following.
4 The coolant level should drop as the engine runs, especially as soon as the thermostat opens, as the coolant is distributed around the various passages of the system. Trapped air will be expelled in the form of bubbles and must be allowed to escape before the radiator cap is refitted. As soon as the temperature gauge has reached its normal level and the level has stabilised with no more air bubbles appearing, stop the engine and top the coolant up to the bottom of the radiator filler neck, then refit and tighten securely the filler cap. If necessary, top the level in the expansion tank up to the 'Full' or 'F' mark.
5 While the coolant level is being checked, carry out the bleeding procedure described in Chapter 3 to remove air bubbles from the oil pump delivery and feed pipes, then refit the oil pump cover. Check the level of oil in the engine oil tank and top up if necessary.
6 It may be necessary to make some adjustment to the carburettor settings as described in Routine maintenance and/or Chapter 3.
7 When all other checks have been completed, stop the engine and check the transmission oil level as described in Routine maintenance; add oil as necessary to achieve the correct level.
8 Finally, refit all remaining components such as air scoops, sidepanels, fairing (where fitted) and the seat.

47 Taking the rebuilt machine on the road

1 Any rebuilt machine will need time to settle down, even if parts

have been refitted in their original order. For this reason it is highly advisable to treat the machine gently for the first few miles to ensure oil has circulated throughout the lubrication system and that new parts fitted have begun to bed down.

2   Even greater care is necessary if the engine has been rebored or if a new crankshaft has been fitted. In the case of a rebore, the engine will have to be run in again, as if the machine were new. This means greater use of the gearbox and a restraining hand on the throttle until at least 500 miles have been covered. There is no point in keeping to any set speed limit; the main requirement is to keep a light loading on the engine and to gradually work up performance until the 500 mile mark is reached. These recommendations can be lessened to an extent when only a new crankshaft is fitted. Experience is the best guide since it is easy to tell when an engine is running freely.

3   Remember that a good seal between the piston and the cylinder barrel is essential for the correct functioning of the engine. A rebored two-stroke engine will require more careful running-in, over a long period, than its four-stroke counterpart. There is a far greater risk of engine seizure during the first hundred miles if the engine is permitted to work hard.

4   If at any time a lubrication failure is suspected, stop the engine immediately and investigate the cause. If an engine is run without oil, even for a short period, irreparable engine damage is inevitable.

5   Do not on any account add oil to the petrol under the mistaken belief that a little extra oil will improve the engine lubrication. Apart from creating excess smoke, the addition of oil will make the mixture much weaker, with the consequent risk of overheating and engine seizure. The oil pump alone should provide full engine lubrication.

6   Do not tamper with the exhaust system. Unwarranted changes in the exhaust system will have a marked effect on engine performance, invariably for the worse. The same advice applies to dispensing with the air filter or its element.

7   When the initial run has been completed allow the engine unit to cool and then check all the fittings and fasteners for security. Re-adjust any controls which may have settled down during initial use and re-check the coolant level in the radiator and the transmission oil level.

Chapter 2 Cooling system

Contents

General description	1
Cooling system: draining	2
Cooling system: flushing	3
Cooling system: filling	4
Radiator: removal and refitting	5
Radiator and cap: cleaning and examination	6
Hoses and connections: examination, removal and refitting	7
Thermostat: removal, testing and refitting	8
Water pump: removal, overhaul and refitting	9
Coolant temperature gauge and sender unit: general	10

Specifications

Coolant
Mixture type .... 50% distilled water, 50% corrosion inhibited ethylene glycol antifreeze

Total capacity of system:
- TZR models .... 1.0 lit (1.7 Imp pt)
- DT models .... 0.92 lit (1.6 Imp pt)

Radiator
- Testing pressure .... 1.0 kg/cm^2 (14.2 psi)
- Cap valve opening pressure .... 0.9 kg/cm^2 (12.8 psi)
- Tolerance .... 0.75 – 1.05 kg/cm^2 (10.7 – 14.9 psi)

Torque settings

Component	kgf m	lbf ft
Coolant drain plug	1.0	7.2
Water pump cover screws	0.8	5.8
Radiator mounting bolts:		
TZR models	0.6	4.3
DT models	0.8	5.8
Radiator cap stopper bolt	0.8	5.8
Thermostat housing screws	0.8	5.8
Cooling system bleed bolt – TZR model	1.0	7.2
Carburettor warmer pipe union bolts	1.2	8.6
Temperature gauge sender unit	1.5	11

1 General description

The liquid cooling system utilises a water/antifreeze coolant to carry away excess energy produced in the form of heat. The cylinder is surrounded by a water jacket from which the heated coolant is circulated by thermo-syphonic action in conjunction with a water pump fitted in the engine right-hand cover and driven via a pinion and shaft from a crankshaft mounted pinion. The hot coolant passes upwards through flexible pipes to the top of the radiator which is mounted on the frame downtubes to take advantage of maximum air flow. The coolant then passes downwards, through the radiator core, where it is cooled by the passing air, and then to the water pump and engine where the cycle is repeated.

The flow of coolant is regulated by a thermostat, a temperature sensitive valve unit contained in a housing on top of the cylinder head. When the engine is cold, the thermostat remains closed, effectively stopping the coolant from circulating through the system. This allows the engine to reach its normal operating temperature rapidly, minimising wear. As the water temperature rises, the thermostat begins to open and the coolant starts to circulate to keep the engine at the optimum temperature.

The hot coolant is also used to warm the carburettor. The coolant flows through a small-bore pipe from the cylinder head, to the carburettor, circulates around the carburettor body and then returns to the main cooling system on the thermostat housing. This prevents the carburettor icing up in extreme weather conditions. The complete system is sealed and pressurised, the pressure being controlled by a valve contained in the spring loaded radiator cap. By pressurising the coolant the boiling point is raised, preventing premature boiling in adverse conditions. The overflow pipe from the radiator is connected to an expansion tank into which excess coolant is discharged by pressure, The expelled coolant automatically returns to the radiator, to provide the correct level when the engine cools again.

2 Cooling system: draining

1 It will be necessary to drain the cooling system on infrequent occasions, either to change the coolant at the recommended interval or to permit engine overhaul or removal. The operation is best undertaken with a cold engine to remove the risk of scalding from hot coolant escaping under pressure.

2 On DT models it will be necessary to remove the left-hand air scoop. This is retained by three screws and can be lifted away once the screws have been removed.
3 Carefully remove the radiator cap stopper bolt. Note that if the engine has been run recently, there will be some residual pressure in the system. If the engine is hot, steam and boiling water may be ejected and can cause scalding. As a precaution, place some rag over the cap and remove it slowly to allow pressure to escape.
4 Place a suitable container under the coolant drain plug situated on the bottom of the water pump housing and remove the drain plug. Tilt the machine slightly to the right to allow the coolant to drain fully.
5 If the system is being drained before an engine overhaul little else need be done at this stage. If the coolant is reasonably new it can be re-used if it is kept clean and uncontaminated. If, however, the system is to be refilled with new coolant it is advisable to give it thorough flushing with tap water, if possible using a hose which can be left running for a while. If the machine has done a fairly high mileage it may be advisable to carry out a more thorough flushing process as described in the following section.

3 Cooling system: flushing

1 After extended service the cooling system will slowly lose efficiency, due to the build-up of scale, deposits from the water and other foreign matter which will adhere to the internal surfaces of the radiator and water channels. This will be particularly so if distilled water has not been used at all times. Removal of the deposits can be carried out easily, using a suitable flushing agent in the following manner.
2 After allowing the cooling system to drain, refit the drain plug and refill the system with clean water and a quantity of flushing agent. Any proprietary flushing agent in either liquid or dry form may be used, providing that it is recommended for use with aluminium engines. Never use a compound suitable for iron engines as it will react violently with the aluminium alloy. The manufacturer of the flushing agent will give instructions as to the quantity to be used.
3 Run the engine for ten minutes at operating temperatures and drain the system. Repeat the procedure twice and then again using only clean cold water. Finally, refill the system as described in the following Section.

4 Cooling system: filling

1 Before filling the system, refit the drain plug using a new sealing washer under its head. Fit and tighten the drain plug to the specified torque setting, and check all the hose clips.
2 Fill the system slowly to reduce the amount of air which will be trapped in the water jacket. Slacken off the bleed bolt on the thermostat housing (TZR models) or slacken the carburettor warmer pipe hose union bolt on the thermostat housing (DT models) until the escaping coolant shows no sign of air bubbles, then tighten the bolt to the specified torque setting. Note that the coolant mixture will attack painted surfaces of the machine if any spilt coolant is not mopped up immediately. When the coolant level is up to the lower edge of the radiator filler neck, run the engine for about 10 minutes at idle speed. Increase engine revolutions for the last 30 seconds to accelerate the rate at which any trapped air is expelled. Stop the engine and replenish the coolant level again to the bottom of the filler neck. Refill the expansion tank up to the 'Full' level mark. Refit the radiator cap, ensuring that it is turned clockwise as far as possible. Refit the radiator cap stopper bolt and tighten to the specified torque.
3 Ideally, distilled water should be used as a basis for the coolant. If this is not readily available, rain water, caught in a non-metallic receptacle, is an adequate substitute as it contains only limited amounts of mineral impurities. Similarly, it is permissible to use tap water which has been thoroughly boiled and allowed to cool down. In emergencies only, tap water can be used, especially if it is known to be of the soft type. Using non-distilled water will inevitably lead to early 'furring-up' of the system and the need for more frequent flushing. The correct water antifreeze mixture is 50/50; do not allow the antifreeze level to fall below 40% as the anti-corrosion properties of the coolant will be reduced to an unacceptable level. Antifreeze of the ethylene-glycol-based type should always be used. Never use alcohol-based antifreeze in the engine.

2.4 Location of coolant drain plug

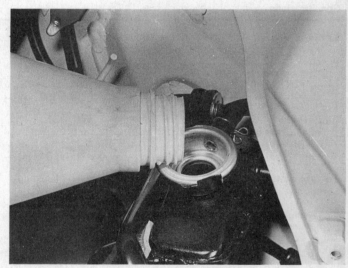
4.2a Mix coolant as described in text and fill the system via the radiator neck – DT model shown

4.2b Refit radiator cap and stopper bolt as shown

5 Radiator: removal and refitting

1 Drain the cooling system as described in Section 2 of this chapter.
2 Disconnect the expansion tank hose from the filler neck and the radiator top and bottom hoses from their respective stubs on the engine.
3 Remove the four radiator grille mounting screws and lift the plastic grille away from the radiator.
4 Slacken and remove both radiator retaining bolts and remove the radiator from the frame. The bolts pass through mounting rubbers; note the order of these so that the radiator can be refitted correctly. The TZR model has an additional mounting at the top which takes the form of a locating peg and locates in a mounting rubber.
5 Refit the radiator by a reversal of the removal sequence making sure that the mounting rubbers are correctly located on reassembly. Refill the cooling system as described in Section 4 of this Chapter.

5.2a Disconnect expansion tank hose from radiator neck ...

5.2b ... followed by top ...

5.2c ... and bottom hoses from engine

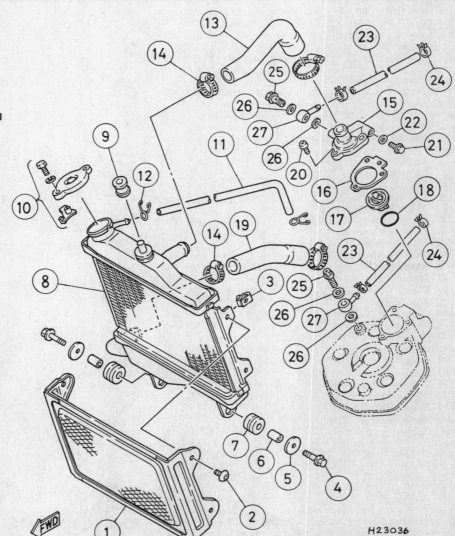

Fig. 2.1 Radiator – TZR model

1 Grille
2 Screw – 4 off
3 Nut – 4 off
4 Bolt – 2 off
5 Washer – 2 off
6 Spacer – 2 off
7 Mounting rubber – 2 off
8 Radiator
9 Mounting rubber
10 Cap assembly
11 Expansion tank pipe
12 Clip – 2 off
13 Top hose (inlet)
14 Hose clamp – 4 off
15 Thermostat housing
16 Gasket
17 Thermostat
18 O-ring
19 Bottom hose (outlet)
20 Screw – 3 off
21 Bleed bolt
22 Sealing washer
23 Carburettor warmer pipe – 2 off
24 Clip – 4 off
25 Union bolt – 2 off
26 Sealing washer – 4 off
27 Pipe union – 2 off

Chapter 2 Cooling system

5.4a TZR model – remove radiator retaining bolts ...

5.4b ... and disengage radiator locating peg to remove radiator

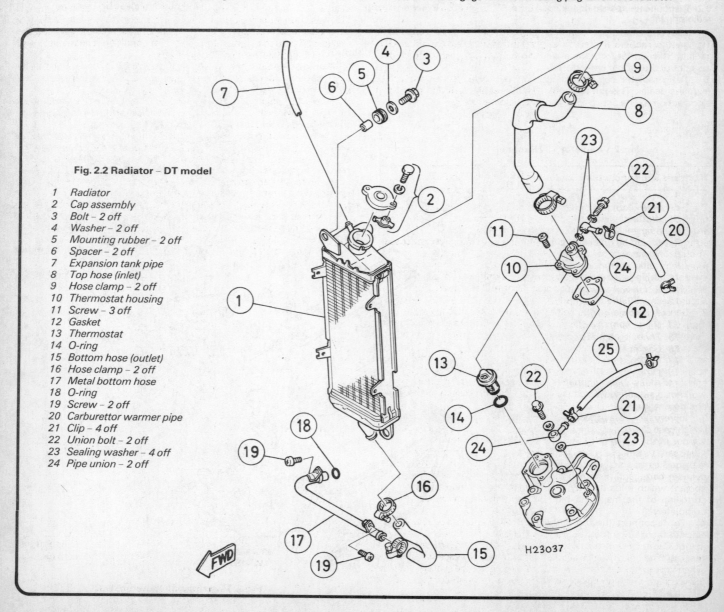

Fig. 2.2 Radiator – DT model

1 Radiator
2 Cap assembly
3 Bolt – 2 off
4 Washer – 2 off
5 Mounting rubber – 2 off
6 Spacer – 2 off
7 Expansion tank pipe
8 Top hose (inlet)
9 Hose clamp – 2 off
10 Thermostat housing
11 Screw – 3 off
12 Gasket
13 Thermostat
14 O-ring
15 Bottom hose (outlet)
16 Hose clamp – 2 off
17 Metal bottom hose
18 O-ring
19 Screw – 2 off
20 Carburettor warmer pipe
21 Clip – 4 off
22 Union bolt – 2 off
23 Sealing washer – 4 off
24 Pipe union – 2 off

6 Radiator and cap: cleaning and examination

1 The interior of the radiator can most easily be cleaned while the radiator is on the motorcycle, using the flushing procedure described in Section 3 of this Chapter. Additional flushing can be carried out by placing the hose in the filler neck and allowing the water to flow through for about ten minutes. Under no circumstances should the hose be connected to the filler neck mechanically as any sudden blockage in the radiator outlet would subject the radiator to the full pressure of the mains supply (about 50 psi). The radiator should not be tested to greater than 1.0 kg/cm² (14.2 psi).

2 Generally, if the radiator is found to be leaking, repair is impracticable and a new component must be fitted. Very small leaks may sometimes be stopped by the addition of a special sealing agent in the coolant. If an agent of this type is used, follow the manufacturer's instructions very carefully. Soldering, using soft solder, may be used for caulking large leaks but this is a specialised repair best left to experts.

3 Remove any obstructions from the exterior of the radiator core, using an air line. The conglomeration of moths, flies and autumnal detritus usually collected in the radiator matrix severely reduces the cooling efficiency of the radiator.

4 If care is exercised, bent fins can be straightened by placing the flat of a screwdriver either side of the fin in question and carefully bending it into its original shape. Badly damaged fins cannot be repaired. If bent or damaged fins obstruct the air flow more than 20%, a new radiator will have to be fitted.

5 Inspect the radiator mounting rubbers for perishing or compaction. Renew the rubbers if there is any doubt as to their condition. The radiator may suffer from the effect of vibration if the isolating characteristics of the rubber are reduced.

6 If the radiator cap is suspect, have it tested by an authorized Yamaha dealer. This job requires specialist equipment and cannot be done at home. The only alternative is to try a new cap.

7 Hoses and connections: examination, removal and refitting

1 The hoses should be inspected periodically and renewed if any sign of cracking or perishing is discovered. The most likely area for this is around the hose clamps which secure each hose to its stub. Particular attention should be given if regular topping up has become necessary. The cooling system can be considered to be a semi-sealed arrangement, the only normal coolant loss being minute amounts through evaporation in the expansion tank. If significant quantities have vanished it must be leaking at some point and the source of the leak should be investigated promptly.

2 Serious leakage will be self-evident, though slight leakage can be more difficult to spot. It is likely that the leak will only be apparent when the engine is running and the system is under pressure, and even then the rate of escape may be such that the hot coolant evaporates as soon as it reaches the atmosphere. Such small leaks may require the use of a special device which will pressurise the system whilst cold and thus enable the leak to be pinpointed. To this end it is best to entrust this work to an authorised Yamaha dealer will have access to the necessary equipment.

3 In very rare cases the leak may be due to a broken head gasket, in which case the coolant may be drawn into the engine and expelled as vapour in the exhaust gases. If this proves to be the case it will be necessary to remove the cylinder head for investigation. If the rate of leakage has been significant if may prove necessary to remove the cylinder barrel and piston so that the crankcase can be checked. Any coolant which finds its way that far into the engine can cause rapid corrosion of the main and big-end bearings and must be removed completely.

4 To disconnect the top and bottom hoses, use a screwdriver to slacken the clamps then slide them along the hose clear of the union spigot. Carefully work the hose off its spigots, rotating it slightly to break the seal. The hoses can be worked off with relative ease when new, or when hot; do not, however attempt to disconnect the system when it is hot as there is a high risk of personal injury through contact with hot components or coolant. The expansion tank pipe is a simple push-fit on the radiator and tank unions, and is secured by a clip at each end.

5 On refitting hoses, first slide the clamps on to the hose and then work it on its respective spigots. **Do not** use lubricant of any type; the hose can be softened by soaking it in boiling water before refitting, although care is obviously required to prevent the risk of personal injury when doing this. When the hose is fitted, rotate it to settle it on its spigots and check that the two components being joined are securely fastened so that the hose is correctly fitted before its clamps are slid into position and tightened securely.

6 The carburettor warmer pipes pass coolant via the union on the cylinder head to the carburettor casting and then back to the thermostat housing via another union. A spring clip at each pipe end secures the pipe to its union or carburettor stub; squeeze the clip ears together to release.

8 Thermostat: removal, testing and refitting

1 The thermostat is automatic in operation and should give many years' service without requiring attention. In the event of a failure, the valve will probably become jammed open, in which case the engine will take much longer than normal to warm up. If, conversely, the valve gets jammed shut, the coolant will be unable to circulate normally and the engine will tend to overheat badly. Neither condition is acceptable, and the fault should be investigated promptly.

2 Remove the sidepanels, seat and fuel tank. On DT models it may be advantageous to remove the exhaust system, although this is not strictly necessary.

3 Drain the cooling system as described in Section 2. Then disconnect both the coolant inlet hose (top hose) and carburettor warmer pipe from the thermostat housing.

4 Remove the three thermostat housing retaining screws and remove the housing. Lift the thermostat and O-ring out of the cylinder head.

5 Examine the thermostat visually before carrying out tests. If it remains in the open position at room temperature, it should be discarded. Suspend the thermostat by a piece of wire in a container of cold water (see accompanying illustration). Place a thermometer in the water so that the bulb is close to the thermostat. Heat the water, noting when the thermostat opens and the temperature at which the thermostat is fully open. If the thermostat is operating correctly it should start to open at 63 – 67°C (146 – 153°F).

6 Refit the thermostat and its housing by reversing the removal

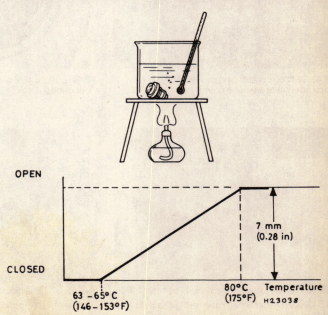

Fig. 2.3 Thermostat operation test

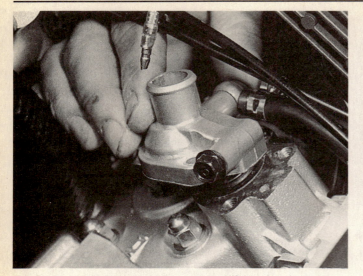
8.4a Remove thermostat housing screws and lift housing clear

8.4b Thermostat can then be removed along with O-ring

sequence. Use a new O-ring if the original is damaged and renew the housing gasket as a matter of course. Fill the cooling system as described in Section 4, then run the engine to check that the thermostat operates normally and that there are no leaks.

9 Water pump: removal, overhaul and refitting

1   The water pump will not normally require attention unless there is obvious leakage of coolant into the transmission oil. To gain access to the pump, drain the coolant and transmission oil fully, then remove the right-hand crankcase cover as described in Section 8 of Chapter 1. Note that if access to the pump cover gasket only is required, there will be no need to remove the crankcase right-hand cover.

2   The water pump is located immediately below the oil pump, the two sharing a common drive pinion on the crankshaft. To dismantle the water pump it will first be necessary to remove its driven gear and washer by displacing the circlip which secures them to the pump spindle. The locating pin should be removed by pushing it through and out of the spindle. Remove the second washer from the spindle.

3   Unscrew the water pump cover screws and remove the cover and

9.4a Drift out water pump seal via hole provided in crankcase cover

9.4b Ensure seal is fitted correctly as described in text

9.4c If necessary drift seal into position using a suitably sized socket

9.5 Grease impeller spindle and oil seal lip before fitting impeller

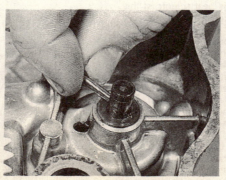
9.6a Refit first plain washer and secure it with locating pin

9.6b Ensure driven gear locates with pin ...

Chapter 2 Cooling system

9.6c ... and refit second plain washer, securing them both with a circlip

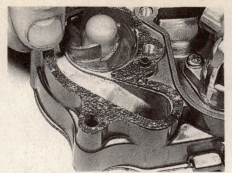

9.6d Check locating dowels are in position and fit a new gasket

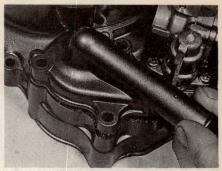

9.6e Refit cover and tighten retaining screws to specified torque setting

gasket. The impeller and spindle may now be displaced and removed.

4 Check the seal for wear or damage and renew if necessary. It can be removed by gently tapping outwards with a suitable drift. Lubricate the outside of the new seal with a light lithium-based grease, and push the seal into place. Note that the seal is marked WATER SIDE on one face, and this should face the water pump impeller. If necessary, the seal can be tapped gently into place; make sure the seal is inserted squarely and is properly seated in the casing.

5 Clean the impeller and spindle, being particularly careful to ensure that any corrosion that may have formed around the seal area is removed and the spindle left completely smooth. If the spindle is badly pitted in this area it may be necessary to renew it to avoid rapid seal wear. The spindle and seal lip should be greased prior to installation and care should be exercised during fitting to avoid damage to the seal face.

6 Reassemble the pump by reversing the removal sequence. Note that the water pump cover gasket should be renewed and that Yamaha specify that the driven gear retaining circlip also be renewed. Tighten all fasteners to the torque settings given in the specifications at the start of this Chapter. Replenish the coolant (Section 4) and transmission oil (Routine maintenance) and check for leaks before taking the machine out on the road.

10 Coolant temperature gauge and sender unit: general

1 Water temperature is monitored by an electrically operated gauge in the instrument panel controlled by a sender unit which screws into the cylinder head water jacket. A description and test procedure of these components will be found in Chapter 7.

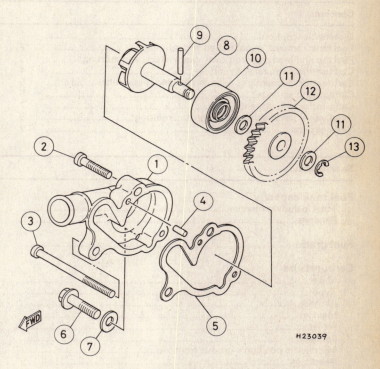

Fig. 2.4 Water pump

1 Pump cover
2 Screw
3 Screw
4 Dowel pin – 2 off
5 Gasket
6 Drain bolt
7 Sealing washer
8 Impeller
9 Locating pin
10 Seal
11 Washer – 2 off
12 Driven gear
13 Circlip

Chapter 3 Fuel system and lubrication

Contents

General description 1	Air filter: general 10
Fuel tank: removal, examination and refitting 2	Exhaust system: general 11
Fuel tap and feed pipe: removal, examination and refitting ... 3	Oil tank: removal and refitting 12
Carburettor: removal and refitting 4	Oil level warning sender unit: removal and refitting 13
Carburettor: dismantling, examination and reassembly 5	Oil pump: removal and refitting 14
Carburettor: settings 6	Oil pump: checking and adjusting the stroke 15
Carburettor: adjustment 7	Oil pump: bleeding 16
Reed valve: removal, examination and refitting 8	Gearbox lubrication: general 17
YEIS: general 9	

Specifications

Fuel tank capacity

	TZR model	DT model
Total including reserve	12 lit (2.6 Imp gal)	10 lit (2.2 Imp gal)
Reserve	1.5 lit (0.3 Imp gal)	1.8 lit (0.4 Imp gal)

Fuel grade .... Unleaded, minimum octane rating 91 (RON/RM)

Carburettors

	TZR model	DT model (3DBI)	DT model (3RN1 onward)
Make	Mikuni	Mikuni	Mikuni
Type	VM26SS	VM26SS	VM26SS
ID mark	2RH00	3BN00	3MB00
Main jet	180	125	210
Air jet	0.8	0.8	0.7
Jet needle	406	407	5J25
Needle clip position – groove from top	4th	3rd	4th
Needle jet	P-6	P-2	Q-2
Throttle valve cutaway	2.0	2.0	2.5
Pilot jet	22.5	25	22.5
Power jet	110	40	60
Air screw (turns out)	1¾	1½	1½
Starter jet	25	25	25

Float height – all models .... 20 – 21 mm (0.787 – 0.827 in)
Fuel level .... see Section 7
Engine idle speed – all models .... 1300 – 1400 rpm

Engine lubrication

Oil tank capacity .... 1.2 litres (0.25 Imp gal)
Oil grade .... Yamaha 2T oil or equivalent good quality air-cooled 2-stroke engine oil

Pump colour code:
- TZR models .... Yellow
- Early DT models (3DB1) .... Red
- Later DT models (3RN1 onward) .... Dark blue

Minimum stroke:
- TZR models and later DT models (3RN1 onward) .... 0.15 – 0.20 mm (0.006 – 0.008 in)
- Early DT models (3DB1) .... 0.20 – 0.25 mm (0.008 – 0.010 in)

Maximum stroke .... 1.85 – 2.05 mm (0.073 – 0.081 in)

Transmission lubrication

Capacity:
- At oil change .... 750 cc (1.3 Imp pint)
- After rebuild .... 800 cc (1.40 Imp pint)
- Recommended oil .... SAE 10W/30 SE motor oil

Chapter 3 Fuel system and lubrication

Reed valve
Valve thickness .... 0.4 mm (0.02 in)
Valve bend limit .... 0.5 mm (0.02 in)
Stopper plate height:
    TZR models .... 8.3 mm (0.327 in)
    DT models .... 6.8 mm (0.268 in)
Tolerance .... ± 0.4 mm (0.02 in)

Torque settings

Component	kgf m	lbf ft
Oil pump mounting screws	0.5	3.5
Carburettor warmer pipe union bolts	1.2	8.6
Reed valve assembly bolts	1.0	7.2
Reed fastening screws	0.1	0.7
Transmission oil drain plug	1.5	11
Exhaust front mounting nuts	1.8	13
Exhaust pipe rear mounting bolt – TZR model	2.5	18
Exhaust pipe rear mounting bolt – DT model	1.0	7.2

1 General description

The fuel system comprises a petrol tank, from which petrol is fed by gravity to the float chamber of the Mikuni carburettor, via a three position petrol tap. The carburettor is of conventional concentric design, the float chamber being integral with the lower part of the carburettor. Cold starting is assisted by a separate starting circuit which supplies the correct fuel-rich mixture when the 'choke' is operated. Later DT models are fitted with a flat slide type carburettor of similar design, except for some minor differences in construction.

Air entering the carburettor passes through an oil-impregnated foam air filter. This effectively removes any airborne dust, which would otherwise enter the engine and cause premature wear. The air cleaner also helps silence induction noise, a common problem inherent with two-stroke engines. The induction system fitted to the machines described in this Manual has two further features; the YEIS chamber, and a reed valve assembly.

Engine lubrication is catered for by the Yamaha Autolube system. Oil from a separate tank is fed by an oil pump to a small injection nozzle in the carburettor body. The pump is linked to the throttle twistgrip, and this controls the volume of oil fed to the engine.

The exhaust system fitted to the TZR models is a single unit comprising the exhaust pipe and expansion chamber/silencer welded together. The DT model exhaust system is similar but incorporates a second, separate silencer assembly and is routed upwards over the engine/gearbox unit to pass down the right-hand side of the machine at a high level in keeping with the trail bike styling of the machine.

2 Fuel tank: removal, examination and refitting

Note: *Petrol is extremely flammable, especially when in the form of vapour. Take all precautions to prevent the risk of fire and read the Safety first! section of this manual before starting work.*

1   Remove both sidepanels and the dualseat. On DT models it will also be necessary to remove the air scoops.

2   Check that the fuel tap is set to the OFF position, then using a suitable pair of pliers, release the wire petrol pipe retaining clip. Slide the clip down the petrol pipe until it is clear of the petrol tap spigot and carefully pull the pipe away from the tap. On TZR models, it will also be necessary to disconnect the breather pipe from the rear of the fuel tank. Slacken and remove the single tank rear mounting bolt, then lift away and put to one side the bolt, washers and damping rubber. Lift the tank up at the rear and ease it backwards off its front mountings, taking care not to damage the paintwork as the mountings are very tight when new and have a nasty habit of releasing their grip so that the tank comes away suddenly.

3   Inspect the tank mounting rubbers for signs of damage or deterioration and, if necessary, renew them before the tank is refitted.

4   In the event that it is necessary to remove the petrol tank for repairs the following points should be noted. Petrol tank repair, whether necessitated by accident or by petrol leaks, is a task for the professional. Welding or brazing is not recommended unless the tank is purged of all petrol vapour; which is a difficult condition to achieve. Resin-based tank sealing compounds are a much more satisfactory method of curing leaks, and are now available through suppliers who advertise regularly in the motorcycle press. Accident damage will inevitably involve re-painting the tank; matching of modern paint finishes, especially metallic ones, is a very difficult task not to be lightly undertaken by the average owner. It is therefore recommended that the tank be removed by the

2.1 DT model – air scoops are retained by three screws

2.2a Check that fuel tap is set to the OFF position ...

Chapter 3 Fuel system and lubrication

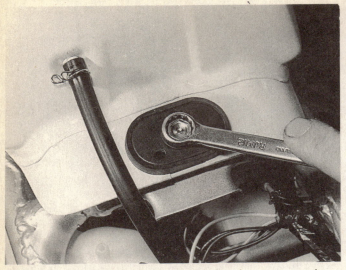

2.2b ... disconnect breather pipe (where fitted) and remove mounting bolt – TZR model shown

2.6 On refitting ensure tank engages correctly with rubber buffers – TZR model shown

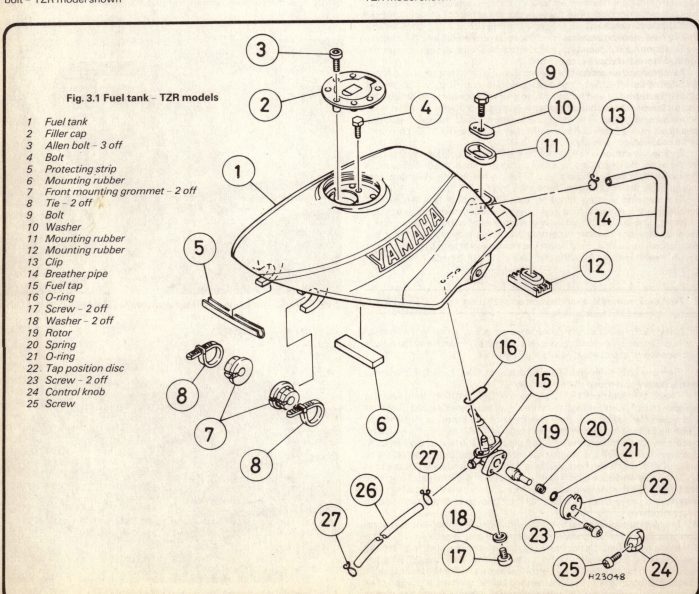

Fig. 3.1 Fuel tank – TZR models

1 Fuel tank
2 Filler cap
3 Allen bolt – 3 off
4 Bolt
5 Protecting strip
6 Mounting rubber
7 Front mounting grommet – 2 off
8 Tie – 2 off
9 Bolt
10 Washer
11 Mounting rubber
12 Mounting rubber
13 Clip
14 Breather pipe
15 Fuel tap
16 O-ring
17 Screw – 2 off
18 Washer – 2 off
19 Rotor
20 Spring
21 O-ring
22 Tap position disc
23 Screw – 2 off
24 Control knob
25 Screw

Chapter 3 Fuel system and lubrication

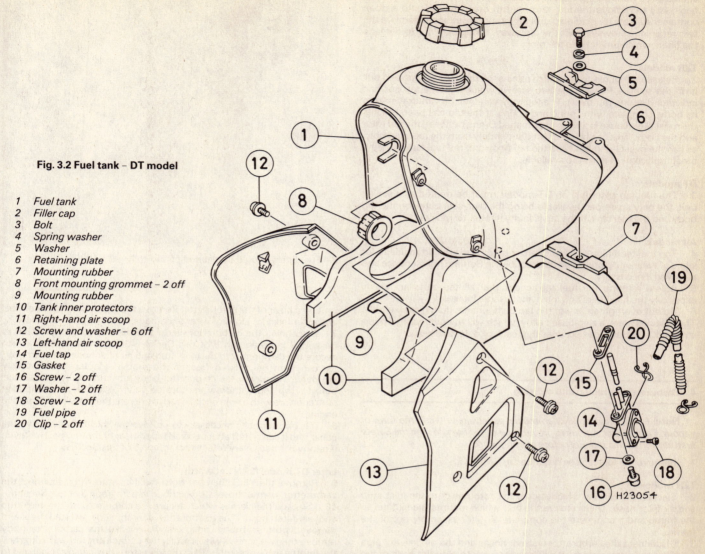

Fig. 3.2 Fuel tank – DT model

1 Fuel tank
2 Filler cap
3 Bolt
4 Spring washer
5 Washer
6 Retaining plate
7 Mounting rubber
8 Front mounting grommet – 2 off
9 Mounting rubber
10 Tank inner protectors
11 Right-hand air scoop
12 Screw and washer – 6 off
13 Left-hand air scoop
14 Fuel tap
15 Gasket
16 Screw – 2 off
17 Washer – 2 off
18 Screw – 2 off
19 Fuel pipe
20 Clip – 2 off

owner, and then taken to a motorcycle dealer or similar expert for professional attention.

5 Repeated contamination of the fuel tap filter and carburettor by water or rust and paint flakes indicates that the tank should be removed for flushing with clean fuel and internal inspection. Rust problems can be cured by using a resin tank sealant.

6 To refit the tank, reverse the procedure adopted for its removal. Move it from side to side before it is fully home, so that the rubber buffers engage with the guide channels correctly. If difficulty is encountered in engaging the front of the tank with the rubber buffers, apply a small amount of grease to the buffers to ease location. Secure the tank with the single retaining bolt whilst ensuring that the mounting components are correctly located and that there is no metal-to-metal contact between the tank and frame.

7 Finally, always carry out a leak check on the fuel pipe connections after fitting the tank and turning the tap lever to the ON position. Any leaks found must be cured; as well as wasting fuel, any petrol dropping onto hot engine castings may well result in fire or explosion.

3 Fuel tap and feed pipe: removal, examination and refitting

Note: *Petrol is extremely flammable, especially when in the form of vapour. Take all precautions to prevent the risk of fire and read the Safety first! section of this manual before starting work.*

1 The fuel tap is retained by two cross-head screws to the underside of the fuel tank. In the event of a leak or other problem, try to work out the most likely area of the fault before removing the tap; fuel seepage

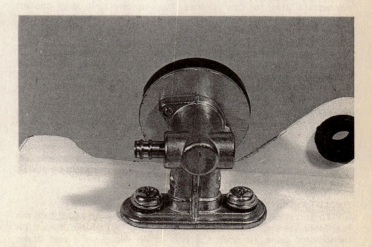

3.1 Fuel tap is retained by two crosshead screws

should be obvious during inspection. Before the tap is removed it will be necessary to remove the fuel tank, and to drain the fuel into a clean container suitable for holding petrol for temporary storage. Remove the two retaining screws and lift the tap away, taking care not to damage the filters which project into the tank.

TZR models
2 Release the single screw which retains the tap control knob and pull it off the spindle. Release the two screws which hold the tap position indicator disc and remove it. The tap rotor can then be withdrawn from its body, complete with O-ring and spring. If the tap has been leaking, the most likely cause is a worn or damaged O-ring between the tap rotor and the body. This can be purchased separately from the tap assembly, as can the oval O-ring between the tap body and the fuel tank, but few other replacement parts are available.

DT models
3 The fuel tap assembly on DT models must be treated as a sealed unit, the only spare part available being the oval seal between the tap body and the fuel tank. If the tap is faulty it must be renewed.

All models
4 The fuel feed pipe and tap filters should be examined regularly for signs of wear or deterioration as described in Routine maintenance and renewed if necessary.
5 Before refitting the fuel tap, check that all the parts are clean, especially the fuel filters and main and reserve intakes.
6 Do not overtighten any of the tap components during reassembly. The castings are in a zinc-based alloy, which will fracture easily if overstressed. Most leakages occur as the result of defective seals.

4 Carburettor: removal and refitting

Note: Petrol is extremely flammable, especially when in the form of vapour. Take all precautions to prevent the risk of fire and read the Safety first! section of this manual before starting work.

1 Remove the fuel tank as described in Section 2.

TZR models
2 Slacken the clamps which secure the carburettor to the inlet stub and air filter hose. Then remove the bolt which mounts the air filter to the frame and manoeuvre the complete air filter assembly out of the frame.
3 Disconnect the carburettor warmer hoses and the oil delivery pipe from the carburettor body. It will be necessary to plug the ends of the pipes to prevent the oil and coolant escaping, so have ready some sort of suitable plug beforehand; a clean screw or bolt of the correct diameter is ideal for this purpose. Mop up any spilt oil or coolant immediately. The carburettor can then be pulled rearwards and removed from the inlet stub. Plug the inlet stub aperture with a wad of clean rag to prevent the entry of dirt into the engine.
4 To release the carburettor body, fully unscrew the threaded carburettor top and remove the throttle valve assembly. It is not normally necessary to remove this from the cable and it can be left attached and taped clear of the engine. However, if removal is necessary proceed as follows:
5 Holding the carburettor top, compress the throttle return spring against it and hold it in position against the top. Invert the throttle valve and shake out the pressed steel spring seat. This component serves to prevent the cable from becoming detached when in position and once out of the way the cable can be pushed down and slid out of its locating groove. The various parts can now be removed and should be placed with the carburettor.

Early DT model (3DB1)
6 To gain access to the carburettor, it is first necessary to detach the rear brake reservoir from its frame mounting and to disconnect the tachometer cable. Remove the bolt which secures the rear brake reservoir to the frame and manoeuvre it clear of the carburettor whilst keeping it upright; if necessary, tie the reservoir to the frame to avoid straining the hydraulic hose. Release its knurled ring and pull the tachometer cable from its drive on the crankcase top, temporarily

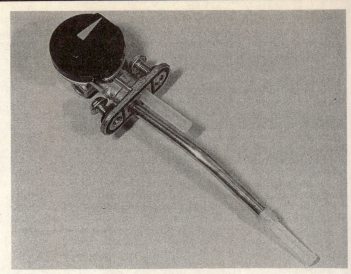

3.4 Fuel filters should be cleaned periodically

routing it clear of the carburettor. Remove the tachometer drive oil seal and examine it for signs of wear or damage, renewing it if necessary.
7 Disconnect the carburettor warmer hoses and oil delivery pipe as described in paragraph 3 of this Section. Slacken the clamps which secure the carburettor to the inlet stub and air filter hose. The carburettor can then be twisted free of the rubber adaptors and partially removed. To release the carburettor body, unscrew the threaded carburettor top and withdraw the throttle valve assembly. Plug the inlet stub aperture with a wad of clean rag to prevent the entry of dirt into the engine.
8 It is not normally necessary to remove the throttle valve from the cable, and it can be left attached and taped clear of the engine. However, if removal is necessary refer to paragraph 5 of this Section.

Later DT model (3RN1 onward)
9 Remove the YEIS chamber from the inlet stub and disconnect the carburettor warmer hoses as described in paragraph 3 of this Section.
10 Slacken the clamps which secure the carburettor to the inlet stub and air filter hose. The carburettor can then be twisted free and removed from its rubber adaptors. To release the carburettor body remove the two screws which retain the carburettor top and withdraw the throttle valve assembly. Plug the inlet stub with a wad of clean rag to prevent the entry of dirt into the engine.

4.2a TZR model – slacken clamps which secure carburettor to inlet stub and air filter hose ...

Chapter 3 Fuel system and lubrication

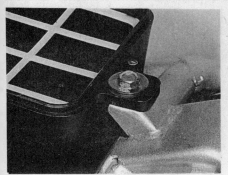
4.2b ... then remove mounting bolt and remove complete air filter assembly from machine

4.3 Disconnect oil delivery pipe and carburettor warmer hoses as described in text

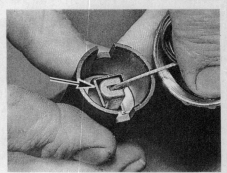
4.5 To remove throttle valve, compress return spring and shake out spring seat (arrowed)

4.10 On flat-slide carburettor the top is retained by two screws

4.12a Take care when refitting throttle valve as needle is easily damaged

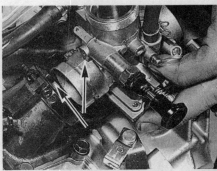
4.12b Ensure tab set in carburettor locates with cutout in inlet stub (arrowed)

11 If required the throttle valve can be separated from the cable as described in paragraph 5 of this Section. If not, it can be left attached to the cable and taped clear of the engine.

All models

12 The carburettor is refitted by reversing the removal sequence. Note that care must be taken when inserting the throttle valve in the carburettor body, to ensure that the jet needle enters the needle jet correctly. Also the groove cut in the throttle valve must correspond with the projection cast in the carburettor body. Check that when fully assembled the throttle valve operates smoothly, without sticking. It is also important that the instrument is mounted vertically to ensure that the fuel level in the float chamber is correct. A locating tab is fitted to provide a good guide to alignment but it is worthwhile checking this for accuracy. If fitted correctly, the tab set in the front of the carburettor body should locate with the cutout in the inlet stub.

13 Once refitted, check the carburettor adjustments as described later in this Chapter. **Note:** If the carburettor is to be set up from scratch it is important to check jet and float height settings prior to installation. To this end, refer to Sections 6 and 7 before the carburettor is refitted.

14 Ensure that the breather and overflow pipes are routed correctly. Reconnect the fuel delivery pipe and secure it with its clip. Turn the fuel tap on and check for leaks. Note that the oil delivery pipe should be bled and the pump adjustment checked after overhaul (see Section 16 and Routine maintenance). Reconnect the carburettor warmer hoses, renewing the union bolt sealing washers if necessary; tighten the union bolts to the specified torque settings. Check the coolant level and top up if required.

5 Carburettor: dismantling, examination and reassembly

Note: *Petrol is extremely flammable, especially when in the form of vapour. Take all precautions to prevent the risk of fire and read the Safety first! section of this manual before starting work.*

1 Remove the carburettor from the machine as described in the previous section.

2 Disconnect the power jet pipe which connects the carburettor body to the float chamber. Invert the carburettor and remove the float chamber by withdrawing the four retaining screws (3 screws on flat-slide carburettor). The float chamber will lift away, exposing the float assembly, hinge and float needle. There is a gasket between the float chamber and the carburettor body which need not be disturbed unless it is leaking. With a pair of thin-nose pliers, withdraw the pin that acts as the hinge for the twin floats. This will free the floats and the float needle. Check that neither of the floats has punctured and that the float needle and seating are both clean and in good condition. If the needle has a ridge, it should be renewed in conjunction with its seating. Examine the float needle's spring-loaded tip; if this is sticking or is damaged in anyway renewal will be necessary. The float needle seat should be removed and examined for any signs of wear. On the flat-slide type carburettor the float needle seat is secured by a small retaining plate and screw, and sealed with an O-ring. On all other models, the seat is threaded into the carburettor body and has a sealing washer under its head. If the float needle and the seat are worn, they should be renewed as a set, never separately. Wear usually takes place in the form of a ridge or groove, which may cause the needle to seat imperfectly.

3 The two floats are made of plastic, connected by a brass bridge and pivot piece. If either float is leaking, it will produce the wrong fuel level in the float chamber, leading to flooding and an over-rich mixture. The floats cannot be repaired successfully, and renewal will be required.

4 Use a slim, narrow-bladed screwdriver to unscrew the pilot jet from its recess to one side of the central jet column. The main jet is located in the centre of the mixing chamber housing. It is threaded into the base of the needle jet and can be unscrewed from the bottom of the carburettor together with its washer. The needle jet can now be removed. The jet is a press fit, and must be pushed out using a short length of wooden dowel, to emerge from the top of the carburettor.

5 Remove the two screws which secure the needle retaining plate to the throttle valve and withdraw the plate and the needle. Check that the needle is straight by rolling it on a flat surface such as a sheet of plate glass and then examine both the needle and needle jet for signs of wear or damage. Any damage to either component will mean that the two

Chapter 3 Fuel system and lubrication

5.2a Disconnect power jet pipe and remove float chamber retaining screws ...

5.2b ... and and lift off float chamber

5.2c Withdraw pivot pin ...

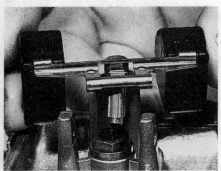
5.2d ... and remove float assembly and needle for examination

5.2e Remove the needle seat as described in text – early type carburettor shown

5.4a Pilot jet can be removed using a slim narrow-bladed screwdriver

5.4b Main jet is a screw fit into needle jet ...

5.4c ... which can then be removed through the top of carburettor

5.5 Needle is retained by a plate inside throttle valve

must be renewed together. Do not attempt to straighten a bent needle as they are easily broken; also, if the machine has been running for any length of time with a bent needle, the needle and needle jet must be renewed anyway to rectify the uneven wear which will have occurred.

6   The needle is suspended from the valve, where it is retained by a circlip. The needle is normally suspended from the groove specified at the front of this Chapter, but other grooves are provided as a means of adjustment so that the mixture strength can be either increased or decreased by raising or lowering the needle. Examine all the O-rings and gaskets in the throttle valve assembly and carburettor top and renew them if necessary. These seals prevent air entering the carburettor which would affect the running of the machine.

7   After an extended period of service the throttle valve will wear and may produce a clicking sound within the carburettor body. Wear will be evident from inspection, usually at the base of the throttle valve and in the locating groove. A worn valve should be renewed as soon as is possible as it will give rise to air leaks which will upset the carburation.

8   The manually operated choke is unlikely to require attention during the normal service life of the machine. The plunger assembly is removed by unscrewing the large brass securing nut. Check that the brass plunger itself is unworn, that the plunger seating is clean and undamaged and that the plunger is operating smoothly. Any faults will mean that the complete assembly must be renewed as repairs are not possible and the assembly is available only as a single unit. Check that the plunger housing and the various passages in the carburettor body are clean and free from any particles of foreign matter.

9   If removal of the throttle stop and pilot adjustment screw is required, screw each one carefully in until it seats lightly, counting and recording the number of turns required, and then unscrew each one, complete with its spring and O-ring. Remove any dirt or corrosion and check for signs of wear or damage, renewing the complete screw assembly if necessary. Refit the screws as shown in the accompanying line drawing. Tighten the screws carefully until they seat lightly, then unscrew each one by the number of turns counted on removal to return

Chapter 3 Fuel system and lubrication

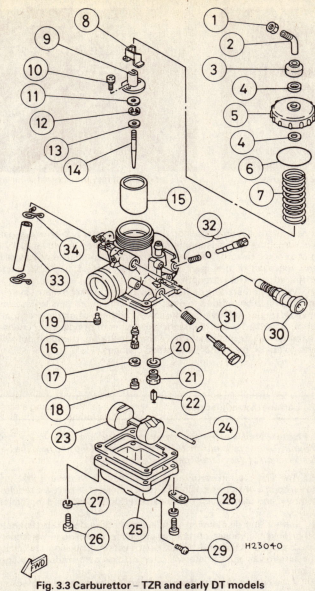

Fig. 3.3 Carburettor – TZR and early DT models

1. Adjuster nut
2. Cable sleeve
3. Cap
4. Sealing washer – 2 off
5. Carburettor top
6. O-ring
7. Return spring
8. Spring seat
9. Needle retaining plate
10. Screw – 2 off
11. Washer
12. Circlip
13. Washer
14. Jet needle
15. Throttle valve
16. Needle jet
17. Washer
18. Main jet
19. Pilot jet
20. Sealing washer
21. Float needle seat
22. Float needle
23. Float
24. Pivot pin
25. Float chamber
26. Screw – 4 off
27. Spring washer – 4 off
28. Pipe guide – 2 off
29. Drain screw
30. Choke control knob
31. Throttle stop screw
32. Pilot screw
33. Power jet pipe
34. Clip – 2 off

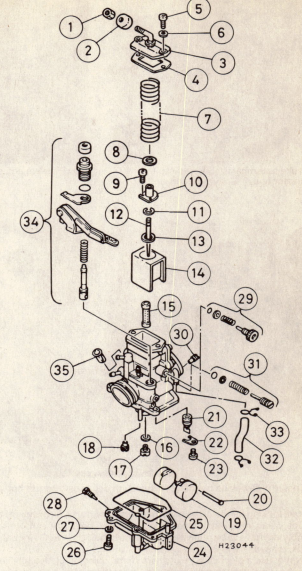

Fig. 3.4 Carburettor – later DT models

1. Adjuster nut
2. Cap
3. Carburettor top
4. Gasket
5. Screw – 2 off
6. Spring washer – 2 off
7. Return spring
8. Spring seat
9. Screw – 2 off
10. Needle retaining plate
11. Circlip
12. Jet needle
13. Washer
14. Throttle valve
15. Needle jet
16. Washer
17. Main jet
18. Pilot jet
19. Float
20. Pivot pin
21. Float needle seat
22. Retaining plate
23. Screw
24. Float chamber
25. O-ring
26. Screw – 3 off
27. Spring washer – 3 off
28. Drain screw
29. Throttle stop screw
30. Air jet
31. Pilot screw
32. Power jet pipe
33. Clip – 2 off
34. Choke control lever
35. Fuel filter

it to its original position; this will serve as a basis for subsequent adjustments.

10 The power jet and starter jet are fixed into the float chamber. Neither jet can be removed but both should be examined to make sure that they are clean and free from obstructions.

11 Before the carburettor is reassembled, using the reversed dismantling procedure, it should be cleaned out thoroughly, preferably by the use of compressed air. Avoid using a rag because there is always risk of fine particles of lint obstructing the internal air passages or the jet orifices. Check carefully the condition of the carburettor body and float

Chapter 3 Fuel system and lubrication

5.8 Choke assembly is screwed into the carburettor body – early type carburettor shown

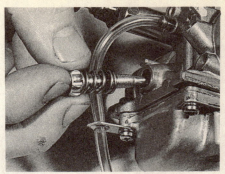
5.9a Remove and examine throttle stop ...

5.9b ... and pilot screw for signs of wear or damage

5.12 Ensure cutaway on needle jet aligns with peg in carburettor body

chamber, looking for distorted or damaged mating surfaces or any other signs of wear. If severe damage or wear is found, the carburettor assembly will have to be renewed. Check the condition of all O-rings and gaskets, renewing any that are worn or distorted.

12 Never use a piece of wire or sharp metal object to clear a blocked jet. It is only too easy to enlarge the jet under these circumstances and increase the rate of petrol consumption. Always use compressed air to clear a blockage; a tyre pump makes an admirable substitute when a compressed air line is not available. Do not use excessive force when reassembling the carburettor because it is quite easy to shear the small jets or some of the smaller screws.

13 Refit the carburettor to the machine as described in the previous Section.

6 Carburettor: settings

1 Some of the carburettor settings, such as the size of the needle jet, main jet and needle position are predetermined by the manufacturer. Under normal circumstances it is unlikely that these settings will require modification, even though there is provision made. If a change appears necessary, if can often be traced to a developing engine fault.

2 As a rough guide, the pilot screw controls the engine speed up to $\frac{1}{8}$ throttle. The throttle valve cutaway controls the engine speed from $\frac{1}{8}$ to $\frac{1}{4}$ throttle and the position of the needle from $\frac{1}{4}$ to $\frac{3}{4}$ throttle. The main jet is responsible for the engine speed at the final $\frac{3}{4}$ to full throttle. It should be added that none of these demarkation lines is clearly defined; there is a certain amount of overlap between the carburettor components involved.

3 Always err on the side of a rich mixture because a weak mixture has a particularly adverse affect on the running of any two-stroke engine. A weak mixture will cause rapid overheating which may eventually promote engine seizure. Reference to Routine maintenance will show how the condition of the spark plug can be used as a reliable guide to carburettor mixture strength.

4 Where non-starting items such as an aftermarket exhaust system or air filter have been fitted to a machine, some alterations to carburation may be required. Arriving at the correct settings often requires trial and error, a method which demands skill borne of previous experience. In many cases the manufacturer of the non-standard equipment will be able to advise on correct carburation changes.

7 Carburettor: adjustment

Note: *Petrol is extremely flammable, especially when in the form of vapour. Take all precautions to prevent the risk of fire and read the Safety first! section of this manual before starting work.*

1 The first step in carburettor adjustment is to ensure that the jet sizes, needle position and float height are correct, which will require the removal and dismantling of the carburettor as described in Sections 4 and 5.

2 Before any dismantling or adjustment is undertaken eliminate all other possible causes of running problems, checking in particular the spark plug, air filter and the exhaust system. Checking and cleaning these items as appropriate will often resolve a mysterious flat spot or misfire.

3 If the carburettor has been removed for the purpose of checking jet sizes, the float height should be measured at the same time. It is unlikely that once this is set up correctly, there will be a significant amount of

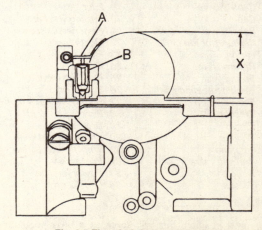

Fig. 3.5 Float height measurement

A Adjusting tang
B Float needle
X Float height – see Specifications

Chapter 3 Fuel system and lubrication

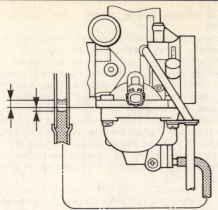

Fig. 3.6 Fuel level measurement

variation, unless the float needle or seat have worn. These should be checked and renewed as required. With the float chamber removed slowly rotate the carburettor until gravity acting on the floats moves the float until the valve is **just** closed, but not so far that the needle's spring-loaded pin is compressed. Measure the distance between the gasket face and the bottom of the float with an accurate ruler. The correct setting should be as given in the Specifications Section. If adjustment is required it can be made by bending, by a very small amount, the small tang to which the float needle is attached. Note that a fuel level is also given. This is measured with the carburettor on the machine and is thus quicker in that respect, but if the level is found to be faulty, the carburettor must still be removed from the machine for the float height to be altered. Since, however, it is arguably more accurate to use the fuel level as a guide, the procedure is given below.

4 The machine must be standing absolutely upright on level ground, as the slightest tilt will make the test inaccurate. Remove the drain/overflow pipe from the spigot at the base of the float chamber and connect it to a length of clear plastic tubing that has an identical internal diameter, then bring this up the front edge of the carburettor body and tape it in place, thus forming a U-tube as shown in the accompanying illustration. Switch the fuel tap on and unscrew the float chamber drain plug by just enough to allow petrol into the tube. Start the engine and allow it to run for a few minutes, to find the correct fuel level, then stop the engine. Measure the distance between the carburettor body bottom edge (body to float chamber mating surface) and the level of fuel in the tube. This should be within the range of −0.5 to +1.5 mm (−0.02 to +0.06 in) on the conventional circular slide type carburettor or −2.5 to +4.5 mm (−0.10 to +0.18 in) on the flat-slide carburettor fitted to the later (3RN1 onward) DT models, see accompanying illustration.

5 Whichever method is used, once the float height/fuel level is known to be accurate and the carburettor is refitted to the machine, start the engine and allow it to warm up to normal operating temperature. Stop the engine and screw the pilot screw in until it seats lightly, then unscrew it by the number of turns shown in the Specifications Section for the particular model. Start the engine and set the machine to its specified idle speed by rotating the throttle stop screw as necessary. Try turning the pilot screw inwards by about $\frac{1}{4}$ turn at a time, noting its effect on the idling speed, then repeat the process, this time turning the screw outwards.

6 The pilot mixture screw should be set in the position which gives the fastest consistent tickover. The tickover speed may be reduced further, if necessary, by unscrewing the throttle stop screw the required amount. Check that the engine does not falter and stop after the throttle twistgrip has been opened and closed a few times.

7 Finally adjust the throttle cable as described in Routine maintenance and check that the throttle valve moves freely before taking the machine on the road. Note that any adjustment of the throttle cable will necessitate that the oil pump cable will also need to be checked and adjusted if necessary, again as described in Routine maintenance.

8 Reed valve: removal, examination and refitting

Note: *Petrol is extremely flammable, especially when in the form of vapour. Take all precautions to prevent the risk of fire and read the Safety first! section of this manual before starting work.*

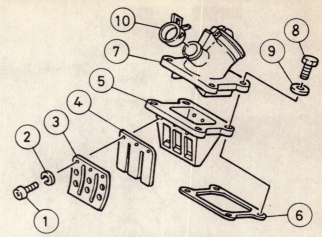

Fig. 3.7 Reed valve assembly

1	Screw – 6 off	6	Gasket
2	Spring washer – 6 off	7	Inlet stub
3	Stopper plate – 2 off	8	Bolt – 4 off
4	Reed petal – 2 off	9	Washer – 4 off
5	Case	10	YEIS hose clip

1 The reed valve assembly is a precision component, and as such should not be dismantled unnecessarily. The valves are located in the inlet tract, covered by the inlet stub.

2 Remove the reed valve assembly as described in Chapter 1, Section 15.

3 The valves can now be washed in clean petrol to facilitate further examination. They should be handled with great care, and on no account dropped. The stainless steel reeds should be inspected for signs of cracking or fatigue, and if suspect, should be renewed. Remember that any part of the assembly which breaks off in service will almost certainly be drawn into the engine, causing extensive damage. The reeds should rest flush against their seats, although a slight gap is quite normal. Check the inlet stub for signs of cracking, perishing or other deterioration and renew if necessary. If further dismantling is necessary proceed as follows:

4 Working on one side at a time to avoid the accidental interchange of components, remove the three cross-head screws securing the valve stopper and reed to the case. Handle the reed carefully to avoid bending. All components should be refitted in their original positions. Lay the reed carefully to one side if it is to be re-used. Examine the neoprene seating face which, if defective, will necessitate the renewal of the complete alloy case, to which it is heat-bonded.

5 Reassembly is a direct reversal of the dismantling process. Clean all parts thoroughly, but gently, before refitting. The reed valve must be installed with its concave face towards the valve case. Note that the manufacturer has provided a small cutout on one corner of the reed valve and valve stopper to assist reassembly; ensure that both cutouts align. A thread-locking compound, such as Loctite, must be applied to the cross-headed screws, which should be tightened progressively to the specified torque setting to avoid warping the reed or stopper. Do not omit the locking compound, as the screws retain a component which vibrates many times each second and consequently are prone to loosening if assembled incorrectly.

6 The assembly should now be checked before refitting. The dimension between the inner edge of the valve stopper and the top edge of the valve case is important as it controls the movement of the reed. If smaller than specified, performance will be impaired. More seriously, if larger than specified, the reed may fracture. The setting is given in the Specifications at the start of this Chapter. Check the valve case mating surfaces for warping, renewing or resurfacing as necessary.

7 Although the reeds should seat flush against the case a small gap is quite normal. This should not however exceed the specified limit, measured between the tip of the reeds and the reed valve case. If the gap is larger than that specified the reeds should be renewed.

8 Refitting the reed valve assembly to the engine is a straightforward reversal of the removal procedure. Always fit new gaskets to prevent induction leaks and be careful not to overtighten the inlet stub bolts;

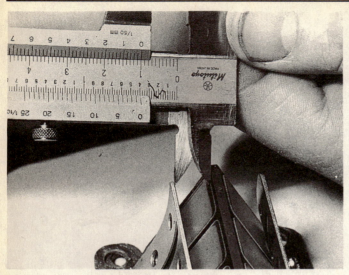

8.6 Measuring reed valve stopper plate height

and ensure that all fasteners are tightened to the specified torque settings.
2   The exhaust must be removed at regular intervals for the carbon deposits to be cleaned out. Refer to Routine maintenance.
3   Do not modify the exhaust system in any way. It is designed to give the maximum power possible consistent with legal requirements and yet to produce the minimum noise level possible. Quite apart from the legal aspects as applied to restricted machines, it is very unlikely that an unskilled person could improve the performance of any of the machines by working on the exhaust. If an aftermarket accessory system is being considered, check very carefully that it will maintain or increase performance when compared with the standard system. Very few 'performance' exhaust systems live up to the claims made by the manufacturers, and even fewer offer any performance increase at all over the standard component.
4   The final point to be borne in mind is the finish. It is inevitable that the original finish will deteriorate to the point where the system must be removed from the machine and repainted, therefore some thought must be given to the type of paint to be used. Several alternative finishes are offered. Reference to the advertisements in the national motorcycle press, or to a local Yamaha dealer and to the owners of machines with similarly-finished exhausts will help in selecting the

note the specified torque setting. When connecting the oil feed pipe, if the end of the pipe is full of oil when it is unplugged, bleeding air from the system will not be necessary. Simply push the end of the pipe over the carburettor union (TZR and early DT models) or inlet stub union (later DT models) and secure it with the clip. If, however, no oil is visible, the feed pipe will have to be bled as described later in this Chapter.

9 YEIS: general

1   The Yamaha Energy Induction System (YEIS), more popularly known as the boost bottle, is a feature developed by Yamaha's design staff to increase intake efficiency in the low to medium speed ranges, thus making the machine easier to ride and increasing fuel economy.
2   The YEIS components are very simple in design and construction having no moving parts at all, and will require no maintenance at all in the life of the machine. The hose and reservoir should be examined whenever the fuel tank is removed during the course of routine maintenance or for repair work. Check carefully for any signs of splits or cracks, and that neither of the components is chafing against anything else. Check very carefully the inlet stub and hose connections for deterioration of any sort, as these are the only two areas likely to suffer. Ensure that the hose clips are securely fastened at all times.
3   If at any time a component is found to be defective, it must be renewed. Use only genuine parts supplied by an authorised Yamaha dealer, as it is essential that the correct internal dimensions and total capacity of the hose and reservoir are preserved if the system is to function correctly.

10 Air filter: general

The care and maintenance of the air filter element is described in Routine maintenance. Never run the engine with the air filter disconnected or the element removed. Apart from the risk of increased engine wear due to unfiltered air being allowed to enter, the carburettor is jetted to compensate for the presence of the filter and a dangerously weak mixture will result if the filter is omitted.

11 Exhaust system: general

1   The exhaust is removed and refitted as described in Sections 4 and 45 of Chapter 1. Always renew the exhaust port gasket to prevent leaks,

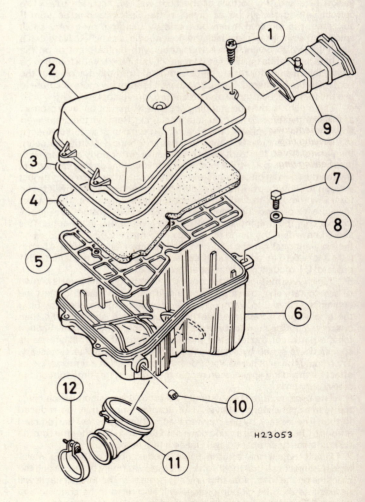

Fig. 3.8 Air filter – TZR models

1	Screw – 5 off	7	Bolt
2	Cover	8	Washer
3	Sealing ring	9	Air intake
4	Element	10	Hose – 2 off
5	Element holder	11	Carburettor adaptor
6	Air filter casing	12	Clamp

Chapter 3 Fuel system and lubrication

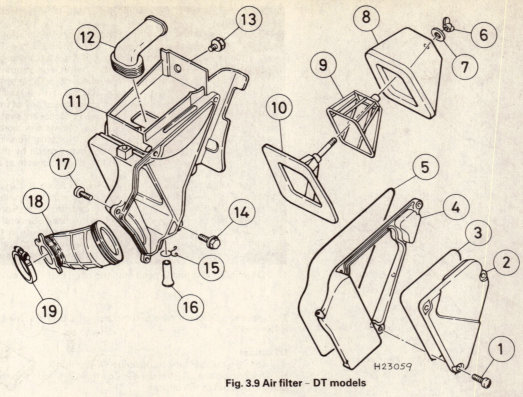

Fig. 3.9 Air filter – DT models

1	Screw – 3 off	6	Wingnut	11	Air filter casing
2	Element cover	7	Washer	12	Air intake
3	Sealing ring	8	Element	13	Bolt
4	Casing cover	9	Element holder	14	Bolt
5	Sealing ring	10	Guide	15	Clip

16	Check hose		
17	Screw – 6 off		
18	Carburettor adaptor		
19	Clamp		

most effective finish. The best are those which require the paint to be baked on, although some high temperature aerosol sprays are almost as effective. Whichever finish is decided upon, ensure that the surface is properly prepared according to the paint manufacturer's instructions and that the paint itself is correctly applied.

5 If any part of the exhaust system is cracked, split or holed through damage or corrosion, it must be renewed immediately to prevent excessive noise and a reduction in performance.

12 Oil tank: removal and refitting

1 The oil tank must be either drained of oil prior to removal or a length of pre-prepared pipe of similar internal diameter, with one end plugged, can be substituted for the 'tank to pump' feed pipe.

TZR model

2 The oil tank is situated under the left-hand sidepanel. To gain access to the tank it will be necessary to remove the seat and sidepanel.
3 Disconnect the sender unit wiring at the block connector. To remove the oil tank, release the two mounting bolts which secure the tank to the frame and manoeuvre it out of the frame. Make a note of how the spring washers, mounting spacers and rubbers are arranged so that they can be correctly positioned when the tank is refitted.

DT model

4 To gain access to the oil tank, remove the three screws which secure the right-hand air scoop to the fuel tank and remove the air scoop. Disconnect the sender unit wiring at the block connector. The oil tank is mounted to the frame by two bolts. Remove both bolts and spacers and lift the oil tank away from the frame.

Fig. 3.10 Exhaust system – TZR models

1	Gasket	6	Sealing washer	
2	Stud – 2 off	7	Bolt	
3	Nut – 2 off	8	Washer	
4	Exhaust system	9	Nut	
5	Bolt	10	Washer	

Chapter 3 Fuel system and lubrication

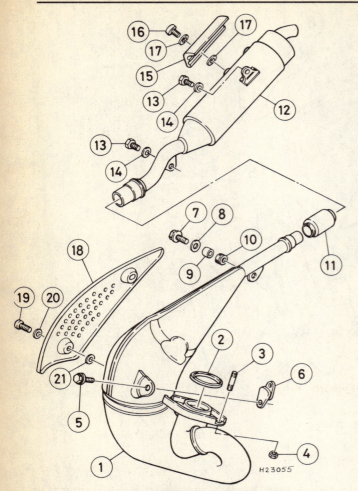

Fig. 3.11 Exhaust system – DT models

1	Exhaust main section	12	Silencer
2	Gasket	13	Bolt – 2 off
3	Stud – 2 off	14	Washer – 2 off
4	Nut – 2 off	15	Heat shield
5	Bolt	16	Screw – 2 off
6	Front mounting bracket	17	Washer – 4 off
7	Bolt	18	Heat shield
8	Washer	19	Screw – 2 off
9	Spacer	20	Washer – 2 off
10	Mounting rubber	21	Washer – 2 off
11	Joining sleeve		

All models
5 The tank is refitted by reversing the removal sequence. Note that to prevent oil starvation on initial start-up, it will be necessary to bleed the oil pump before use. Refer to Section 16 of this Chapter.

13 Oil level warning sender unit: removal and refitting

1 The low oil level warning system, comprises the tank-mounted sender unit, the warning lamp in the instrument panel and the relevant wiring.
2 The sender unit is a push fit in the top of the oil tank. To remove the sender unit, proceed as follows.

TZR model
3 Remove the seat and left-hand sidepanel and release the oil tank from its mountings as described in the previous Section. Note that it is not necessary to disconnect the oil feed pipe. Unplug the oil level sender unit from the main wiring loom and pull the unit out of the oil tank.

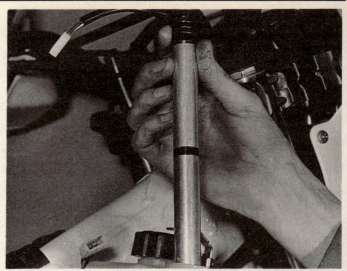

13.2 Oil level sender unit is push fit in top of oil tank – DT model shown

Temporarily refit the oil tank and mounting bolts whilst the sender unit is removed.

DT model
4 Remove the fuel tank as described in Section 2.
5 Disconnect the oil level sender unit from the main wiring loom and pull the unit out of the oil tank.

All models
6 Refer to Chapter 7 for details of the sender unit test procedure.

14 Oil pump: removal and refitting

1 It is rarely necessary to remove the oil pump unless specific attention to it is required. It should be noted that the pump should be considered a sealed unit – parts are not available and thus it is not practicable to repair it. The pump itself can be removed quite easily leaving the driveshaft and pinion in place in the right-hand outer casing. If these latter components require attention it will be necessary to drain the cooling system and transmission oil so that the right-hand outer casing can be removed. Refer to Chapter 1 for further details.
2 To gain access to the oil pump, remove the screws which secure the pump cover to the right-hand outer casing.
3 Displace the small spring steel clip, which secures the oil delivery pipe to the pump outlet, then ease the pipe off the outlet stub using a small screwdriver. The large feed pipe from the oil tank is removed in a similar fashion, but before removing it have some sort of plug handy (such as a clean screw of suitable size) to push into the end of the pipe. This will prevent the oil from the tank being lost. Pull on the pump cable inner to rotate the pump pulley. Holding the pulley in its fully open position release the cable and disengage it from the pulley recess. On DT models there is a small spring clip which locates the pump cable to the pulley; this must be removed before the cable can be freed.
4 The pump is secured to the casing by two screws which pass through its mounting flange. Once these have been removed the pump can be removed, noting that it may prove necessary to turn the pump slightly to free it from its driveshaft.
5 Further dismantling is not practicable, and it will be necessary to renew the pump if it is obviously damaged. Maintenance must be confined to keeping the pump clear of air, and correctly adjusted, as described in Section 16 and Routine maintenance respectively.
6 Refit the oil pump to the crankcase cover, using a new gasket at the oil pump/crankcase cover joint. Replace and tighten the two crosshead mounting screws to the specified torque. The remainder of the reassembly is accomplished by reversing the dismantling procedure, but do not replace the pump cover at this stage because the oil pump must be

Chapter 3 Fuel system and lubrication

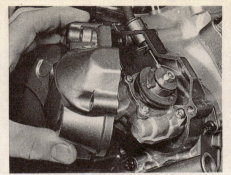
14.2 Remove retaining screws and lift oil pump cover away

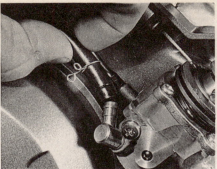
14.3a Disconnect oil delivery and feed pipes ...

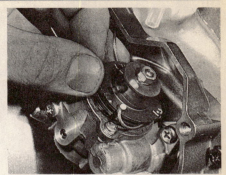
14.3b ... and pump control cable (see text)

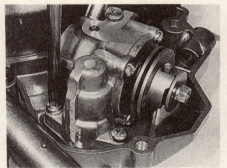
14.4 Pump is retained by two crosshead screws

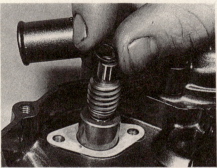
14.6 Do not omit shim when refitting oil pump and use a new gasket

14.9a Insert drive shaft ...

14.9b ... and secure it with a circlip as shown

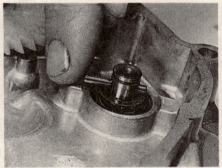
14.9c Refit locating pin ...

14.9d ... and nylon drive gear – secure with the second circlip

bled to ensure the oil lines are completely free from air bubbles. See Section 16. When bleeding of the system is complete check the minimum stroke setting (Section 15) and cable adjustment (Routine maintenance).

7 Should it be necessary at any time to inspect the oil pump drive components, the crankcase right-hand cover must be removed as described in Section 8 of Chapter 1. Remove the oil pump and invert the cover. Using a small-bladed screwdriver, prise the driven gear retaining circlip out of its groove in the pump driveshaft and lift the white nylon driven gear away, so that the gear locating pin and the second circlip can be withdrawn. Push the shaft out of its bush in the cover.

8 As the oil pump drive components are lightly loaded and well lubricated, wear is unlikely to occur until a very high mileage has been covered. Damage of any sort should be visible easily and must be rectified by the renewal of the parts concerned. The shaft oil seal may be prised from its housing with a suitably shaped screwdriver. If necessary the driveshaft bush may be driven out using a hammer and a drift of suitable size, after the casing has been heated by immersing in boiling water. The state of wear of the driveshaft and bush can be assessed by feeling for excessive free play when the shaft is installed in the bush. Renew the shaft or the bush as necessary.

9 The oil pump drive components are refitted by a reversal of the removal sequence.

15 Oil pump: checking and adjusting the stroke

1 The minimum stroke setting should not alter readily, but should be checked if there is some question regarding the rate of oil delivery to the engine. Remove the oil pump cover, start the engine and allow it to idle. Observe the front of the pump where it will be noticed that the pump adjustment plate moves in and out. Wait until the plate is fully out and stop the engine. Using feeler gauges, measure the gap between the plate and the raised boss of the pulley, making a note of the reading. Start the engine and repeat the check several times, taking the largest gap as the minimum stroke position. This should be within the specified

Chapter 3 Fuel system and lubrication

Fig. 3.12 Oil pump

1 Oil pump
2 Screw – 2 off
3 Gasket
4 Washer
5 Drive shaft
6 Drive pin
7 Bush
8 Oil seal
9 Circlip – 2 off
10 Drive gear
11 Nut
12 Spring washer
13 Adjusting plate
14 Shim – as required
15 Spring clip △
16 Bleed screw △
17 Sealing washer △
18 Bleed screw ▲
19 Sealing washer ▲
20 Oil feed pipe (from tank)
21 Clip
22 Steel ball ▲
23 Spring ▲
24 Union ▲
25 Spring clip
26 Oil delivery pipe (to engine)
▲ TZR models
△ DT models

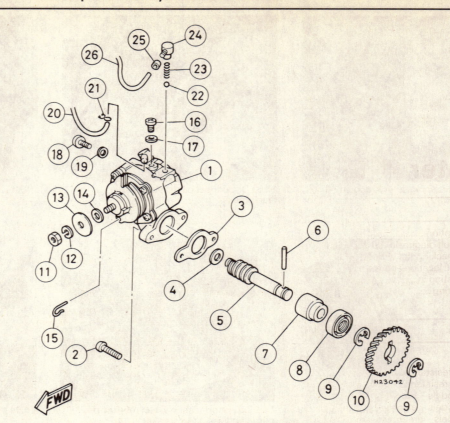

range given in the Specifications at the start of this Chapter. If not, remove the adjuster plate locking nut and spring washer, adjusting plate and shim(s). Add or subtract shims to bring the setting within the specified limits; shims can be purchased from an authorised Yamaha dealer. Refit the adjusting plate, spring washer and locknut and recheck the minimum stroke, repeating the above sequence until the pump is within the specified range.

2 Check the oil pump cable adjustment, as described in Routine maintenance, and refit the cover.

16 Oil pump: bleeding

1 It is necessary to bleed the oil pump every time the main feed pipe from the oil tank is disturbed. This is because air will be trapped in the oil pipe, no matter what care is taken when the pipe is removed.

2 Check that the oil feed pipe is connected correctly, with the retaining clip in position. Then remove the cross-head screw in the outer or upper face of the pump body with the fibre washer beneath the head. This is the oil bleed screw.

3 Check that the oil tank is topped up to the correct level, then place a container below the oil bleed hole to collect the oil that is expelled as the pump is bled. Allow the oil to trickle out of the bleed hole, checking for air bubbles. The bubbles should eventually disappear as the air is displaced by fresh oil. When clear of air, refit the bleed screw. Do not replace the front portion of the crankcase cover until the pump cable setting has been checked, as described in Routine maintenance.

4 Note also that it will be necessary to ensure that the oil delivery pipe is primed if it has been disturbed. Unless this is checked the engine will be starved of oil until the pipe fills. The procedure required to avoid this is to start the engine and allow it to idle for a few minutes whilst holding the pump pulley in its fully open position by pulling the pump cable. The

16.2 Bleeding the oil pump – TZR model shown

excess oil will make the exhaust smoke heavily for a while, indicating that the pump is delivering oil to the engine.

17 Gearbox lubrication: general

The gearbox, clutch and primary drive are lubricated by a splash feed from an oil bath in the crankcases. Full details of checking the oil level and carrying out an oil change can be found in Routine maintenance.

Chapter 4 Ignition system

Contents

General description	1
CDI system: fault diagnosis	2
CDI system: checking the wiring	3
Ignition HT coil: location and testing	4
Source coil: testing	5
Pulser coil: testing	6
Ignition and engine kill switch: testing	7
CDI unit: testing	8
Ignition control system: testing	9
Ignition timing: general	10
HT lead and suppressor cap: examination	11
Spark plug: general	12

Specifications

Generator
Pulser coil resistance .... 280 – 420 ohm @ 20°C (68°F)
Source coil resistance:
    TZR models – brown to yellow .... 48 – 72 ohm @ 20°C (68°F)
    TZR models – black/red to black .... 520 – 780 ohm @ 20°C (68°F)
    DT models – green/white to black/red .... 192 – 288 ohm @ 20°C (68°F)

Ignition timing (BTDC)
TZR models .... 16° @ 1350 rpm
Early DT models (3DB1) .... 19° @ 1350 rpm
Later DT models (3RN1 onward) .... 17° @ 1350 rpm

Ignition HT coil
Minimum spark gap .... 6 mm (0.24 in)
Primary winding resistance .... 0.7 – 1.1 ohm @ 20°C (68°F)
Secondary winding resistance .... 5.7 – 8.5 K ohm @ 20°C (68°F)

Spark plug
Type: **NGK**   **Nippon Denso**
    TZR models .... BR8ES or BR9ES   W24ESR-U or W27ESR-U
    DT models BR9ES   W27ESR-U
Gap across electrodes .... 0.7 – 0.8 mm (0.028 – 0.031 in)
Suppressor cap resistance .... 4.0 – 6.0 K ohm @ 20° (68°F)

1 General description

The machines described in this Manual are equipped with fully electronic capacitor discharge ignition (CDI) systems. This arrangement provides a more powerful and accurate ignition system and can be considered maintenance-free.

TZR model
Energy for the ignition system is drawn from the source coil. This is mounted on the generator stator, and is integral with the normal alternator windings. It is a two-stage arrangement, having low speed and high speed windings. The low speed windings produce a high output voltage at low engine speeds, this voltage dropping off as the engine builds up speed. The high speed windings, on the other hand, produce little energy at low engine speed, but the output voltage rises along with engine speed. The two outputs are combined, offsetting each other to give a fairly constant output voltage, this being the sum of the output of each set of windings.

The source coil assembly feeds the CDI unit, a sealed electronic assembly which forms the heart of the system. This unit contains, amongst other things, a capacitor and a thyristor, or silicon controlled rectifier (SCR). The capacitor is charged with the high voltage output from the source coil assembly. The thyristor, or SCR, is in effect an electronic switch. When signalled electrically by the pulser coil, it allows the capacitor to discharge through the primary windings of the ignition coil. This in turn induces a high tension pulse in the secondary windings, which is fed to the spark plug.

The pulser, or pickup, comprises a small coil mounted outside the generator rotor on a projection from the stator. A permanent magnet embedded in the flywheel rotor is arranged to pass close to the pulser coil. As the magnet passes the pulser coil, a weak current is generated and it is this that is used to trigger the thyristor in the CDI unit.

DT model
The DT model ignition system operates on the same principles as the TZR model but differs in that only one source coil is fitted. The coil is not an integral part of the alternator stator and can be renewed as a separate item if necessary.

All models
Both machines are equipped with an ignition control system, incorporated as a safety feature. The ignition control system circuit consists of the ignition control unit, neutral switch and sidestand switch, al-

Chapter 4 Ignition system

though on later models the control unit is an integral part of the CDI unit. The control system will automatically earth the current from the source coil if the transmission is put into gear with the sidestand down. This effectively breaks the ignition system and kills the engine.

2 CDI system: fault diagnosis

1 As no means of adjustment is available, any failure of the system can only be traced to the failure of a system component or a simple wiring fault. Of the two possibilities, the latter is by far the most likely. In the event of failure, check the system in a logical fashion, as described below.
2 Remove the spark plug, giving it a quick visual check noting any obvious signs of flooding or oiling. Fit the plug into the plug cap and rest it on the cylinder head so that the metal body of the plug is in good contact with the cylinder head metal. The electrode end of the plug should be positioned so that sparking can be checked as the engine is spun over using the kickstart.
3 **Important note** *The energy levels in electronic systems can be very high.* **On no account** *should the ignition be switched on whilst the plug or plug cap is being held. Shocks from the HT circuit can be most unpleasant. Secondly, it is vital that the plug is soundly earthed when the system is checked for sparking. The CDI unit* **can be seriously damaged** *if the HT circuit becomes isolated.*
4 Having observed the above precautions, turn the ignition switch to 'On' and kick the engine over. If the system is in good condition a regular, fat blue spark should be evident at the plug electrodes. If the spark appears thin or yellowish, or is non-existent, first substitute a brand new plug of the specified make and type and repeat the test. If this fails to work try a new suppressor cap. If there is still no spark, further investigation will be required. Before proceeding further, turn the ignition off and remove the key as a safety measure.
5 Ignition faults can be divided into two categories, namely those where the ignition system has failed completely, and those which are due to a partial failure. The likely faults are listed below, starting with the most probable source of failure. Work through the list systematically, referring to the subsequent sections for full details of the necessary checks and tests.
  (a) Loose, corroded or damaged wiring connections, broken or shorted wiring between any of the component parts of the ignition system
  (b) Faulty ignition HT coil
  (c) Faulty source coil
  (d) Faulty pulser coil
  (e) Faulty ignition switch
  (f) Faulty ignition control system
  (g) Faulty CDI unit

3 CDI system: checking the wiring

1 The wiring should be checked visually, noting any signs of corrosion around the various terminals and connectors. If the fault has developed in wet conditions it follows that water may have entered any of the connectors or switches, causing a short circuit. A temporary cure can be effected by spraying the relevant area with one of the proprietary de-watering aerosols such as WD40 or similar. A more permanent solution is to dismantle the switch or connector and coat the exposed parts with silicone grease to prevent the ingress of water. The exposed backs of connectors can be sealed off using a silicone rubber sealant.
2 Light corrosion can normally be cured by scraping or sanding the affected area, though in serious cases it may prove necessary to renew the switch or connector affected. Check the wiring for chafing or breakage, particularly where it passes close to part of the frame or its fittings. As a temporary measure damaged insulation can be repaired with PVC tape, but the wire concerned should be renewed at the earliest opportunity.
3 Using the wiring diagram at the end of the manual, check each wire for breakage or short circuits using a multimeter set on the resistance scale or a dry battery and bulb wired as shown in Fig. 7.1. In each case, there should be continuity between the ends of each wire.

4 Ignition HT coil: location and testing

1 If the ignition coil is suspected of having failed it can be tested by measuring the resistance of its primary and secondary windings. The test can be performed with the coil in place on the frame, having first disconnected the HT lead at the spark plug and the orange lead from the ignition coil. The ignition HT coil is located under the fuel tank where it is mounted between the frame tubes.
2 Set the meter to the ohms x 1 scale, connect one of the meter probes to the earth on the coil's centre pole and the other to the low tension terminal. This will give a resistance reading of the primary windings and should be within the limits given in the Specifications.
3 Unscrew the suppressor cap from the end of the HT lead. Set the meter to the K ohm scale, connect one of the meter probes to the HT lead and the other to earth on the coil's centre pole. This will measure the secondary winding resistance and should compare with the value given in the Specifications.
4 If either of the values obtained differs markedly from the specified resistances it is likely that the coil is defective. It is recommended that the suspect coil is taken to an authorized Yamaha dealer who can then verify the coil's condition using a spark gap tester. The coil is a sealed unit and therefore cannot be repaired.

5 Source coil: testing

1 Trace the generator output wiring back to the connector blocks and disconnect the generator wiring from the main wiring harness. Make the tests on the generator side of the connector.

4.1 Ignition HT coil location – TZR model

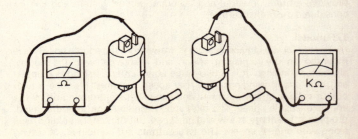

Primary Winding   Secondary Winding

Fig. 4.1 Ignition HT coil windings test

Chapter 4 Ignition system

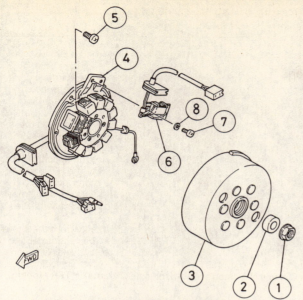

Fig. 4.2 Flywheel generator – TZR models

1. Nut
2. Washer
3. Rotor
4. Stator plate and coils
5. Screw – 2 off
6. Pulser coil
7. Screw – 2 off
8. Spring washer – 2 off

TZR model

2 Set the meter to the ohms x 10 scale, measure the resistance between the brown and yellow leads (high speed windings) and note the reading. Next, with the meter set on the ohms x 100 scale, measure the resistance between the black/red and black leads (low speed windings). The readings should compare with those given in the Specifications.

DT model

3 Set the meter to the ohms x 100 scale and measure the resistance between the green/white and black/red leads. Compare the reading obtained with that given in the Specifications.

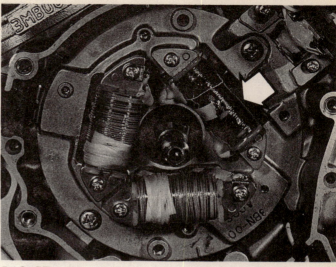

5.4 On DT model, generator coils can be renewed individually – source coil arrowed

All models

4 If the results obtained differ widely from those specified the coil must be renewed. Check first, however, that the fault is not due to a broken wire or connection. This will necessitate the removal of the generator rotor as described in Chapter 1, Section 14. In the case of the DT model the source coil can be obtained as a separate part, although on TZR models it will be necessary to purchase the complete generator stator.

6 Pulser coil: testing

1 Trace the generator output wiring back to the connector blocks and disconnect the generator wiring from the main wiring harness. Make the following test on the generator side of the wiring.

2 Identify the two pulser coil wires, white/red and white/green for the TZR models, and white/blue and white/red for DT models. Measure the resistance between them using a multimeter set on the ohm x 100 scale. The pulser coil must be renewed if the reading obtained differs greatly

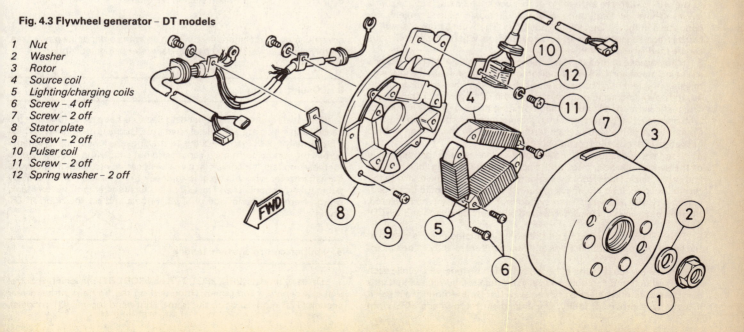

Fig. 4.3 Flywheel generator – DT models

1. Nut
2. Washer
3. Rotor
4. Source coil
5. Lighting/charging coils
6. Screw – 4 off
7. Screw – 2 off
8. Stator plate
9. Screw – 2 off
10. Pulser coil
11. Screw – 2 off
12. Spring washer – 2 off

6.1 Pulser coil location

8.1a CDI unit is located on left-hand side of frame on TZR model ...

8.1b ... and on the right-hand side on DT model

from that given in the Specifications, particularly if the meter indicates a short-circuit (no measurable resistance) or an open circuit (infinite, or very high resistance).

3 If a coil is thought to be faulty, first check that this is not due to a damaged or broken wire from the coil itself to the connector; pinched or broken generator lead wires can usually be repaired by the average private owner. If the pulser is confirmed faulty, the flywheel generator rotor must be removed as described in Chapter 1, Section 14 and the coil renewed.

7 Ignition and engine kill switches: testing

1 The ignition system is controlled by the ignition switch or main switch which is bolted to the top fork yoke. The switch has several terminals, of which two are involved in controlling the ignition system. These are the separate black/white lead and the black lead in the connector block. The two terminals are connected when the switch is in the 'Off' position and prevent the ignition system from functioning by shorting the CDI unit to earth. When the switch is in the 'On' position the CDI/earth connection is broken and the system is allowed to function.

2 If the operation of the switch is suspect, reference should be made to the wiring diagram at the end of this book. The switch connections are also shown in diagrammatic form and indicate which terminals are connected in the various switch positions. The wiring from the switch can be traced back to the respective connectors where test connections can be made most conveniently.

3 The purpose of the test is to check whether the switch connections are being made and broken as indicated by the diagram. In the interests of safety the test must be made with the machine's battery disconnected, thus avoiding accidental damage to the CDI system or the owner. The test can be made with a multimeter set on the resistance scale, or with a simple dry battery and bulb arrangement as shown ion Fig. 7.1. Connect one probe lead to each terminal and note the reading or bulb indication in each switch position.

4 If the test indicates that the black/white lead is earthed irrespective of the switch position, trace and disconnect the ignition (black/white) and earth (black) leads from the ignition switch. Repeat the test with the switch isolated. If no change is apparent, the switch should be considered faulty and renewed. If the switch works normally when isolated, the fault must lie in the black/white lead between the switch and the CDI unit.

5 The kill switch, mounted on the right-hand handlebar, is tested in the same way as its connections and functions are exactly as described above.

6 If either switch is found to be faulty it must be renewed. While each is a sealed unit and can only, officially, be repaired by renewing it as a complete assembly, there is nothing to be lost by attempting to repair it if tests have proven it faulty. Depending on the owner's skill, worn contacts may be reclaimed by building up with solder or in some cases, merely cleaning with WD40 or a similar water dispersant spray.

8 CDI unit: testing

1 If the tests shown in the preceding Sections have failed to isolate the cause of an ignition fault it is likely that the CDI unit is itself faulty. Whilst it is normally possible to check this by making resistance measurements across the various terminals, Yamaha do not supply the necessary data for these models. It follows that it will be necessary to enlist the help of a Yamaha dealer who will be able to check the operation of the unit by substituting a sound item. The CDI unit is located beneath the fuel tank, on the frame left-hand side on TZR models and on the right on DT models.

9 Ignition control system: testing

1 On early models (2RK, 3PC1, TZR and 3DB1 DT) the ignition control unit is a separate component, mounted on the frame right-hand rear section (TZR models) or on the frame right-hand top rail (DT models).

Later models are fitted with a combined CDI unit and ignition control unit.

2 On early models the test described below enables the ignition control circuit to be isolated for testing. On later models, however, this is not possible; if a fault is suspected in this circuit all that can be done is to check the side stand and neutral switches for continuity as described in paragraphs 4 and 5 below. If both switches are functioning correctly and the wiring is sound it is likely that the fault lies in the combined CDI/ignition control unit, which can only be checked by the substitution of a known sound component. Check first that the fault does not lie elsewhere in the main ignition circuit before condemning the CDI/ignition control unit.

3 On early models disconnect the brown wire on the ignition control unit, isolating the ignition control unit, and turn the ignition on. Making sure the kill switch is set to the run position, attempt to kickstart the engine. If the engine starts, the fault lies somewhere in the ignition control system and the system should be checked using the following procedure.

4 Trace the sidestand switch wiring back to its connectors and separate them from the main wiring harness. Set the multimeter to the ohms x 1 scale, and check the switch for continuity between the blue/yellow and black leads. Continuity should only be present when the sidestand is in the up position, if not the switch must be renewed.

5 Test the neutral switch using the same technique. Trace the generator output wiring back to the connectors and separate it from the main wiring harness. Check for continuity between the light blue lead from the switch and the frame. Continuity should only be present when the transmission is in neutral, if not the switch must be renewed.

6 Check the wiring connections as described in Section 3. Then reconnect the brown wire on the ignition control unit and attempt to kickstart the engine.

7 If the preceding tests have failed to isolate the problem, and the engine still will not start, the ignition control unit itself must be faulty. It is not possible to test or repair the unit so a new unit must be purchased.

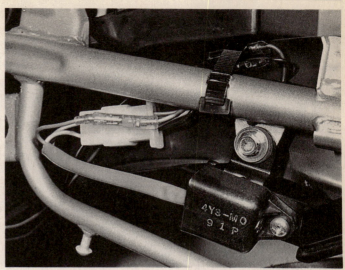

9.1 Ignition control unit location – TZR 2RK and 3PC1 models

10 Ignition timing: general

The ignition timing is fixed and is thus unlikely to go out of adjustment. The only possible cause for inaccurate ignition timing is an internal fault in the CDI unit or a full or partial failure of the pulser coil windings.

11 HT lead and suppressor cap: examination

1 Erratic running faults and problems with the engine suddenly cutting out in wet weather can often be attributed to leakage from the high tension lead and suppressor cap. If this fault is present, it will often be possible to see tiny sparks around the lead and cap at night. One cause of this problem is the accumulation of mud and road grime around the lead, and the first thing to check is that the lead and cap are clean. It is possible to cure the problem by cleaning the components and sealing them with an aerosol ignition sealer, which will leave an insulating coating on both components.

2 Water dispersant sprays are also highly recommended where the system has become swamped with water. Both these products are easily obtainable at most garages and accessory shops. Occasionally, the suppressor cap or the lead itself may break down internally. If this is suspected, the components should be renewed.

3 Where the HT lead is permanently attached to the ignition coil it is recommended that the renewal of the HT lead is entrusted to an auto-electrician will have the expertise to solder on a new lead without damaging the coil windings.

4 When renewing the suppressor cap, be careful to purchase one that is suitable for use with resistor spark plugs.

12 Spark plug: general

If the spark plug is thought to be faulty, the only possible means of testing is to renew the plug. Refer to Routine maintenance for further information.

Chapter 5 Frame and forks

Contents

General description	1
Front forks: removal	2
Front forks: dismantling and reassembly	3
Front forks: examination and renovation	4
Front forks: refitting	5
Steering head: dismantling	6
Steering head bearings: removal, examination and refitting	7
Steering head: reassembly	8
Frame: examination and renovation	9
Swinging arm and suspension linkage: removal	10
Swinging arm and suspension linkage: examination and renovation	11
Swinging arm and suspension linkage: refitting	12
Rear suspension unit: removal	13
Rear suspension unit: examination and renovation	14
Rear suspension unit: refitting and adjustment	15
Footrests, stands and controls: examination and renovation	16
Instrument panel: removal and refitting	17
Instrument drive cables: examination and maintenance	18
Instrument drives: examination and maintenance	19
Fairing: removal and refitting – TZR model	20
Seat and sidepanels: removal and refitting	21

Specifications

Front forks

	TZR model	DT model (3DB1)	DT models (3RN1-on)
Wheel travel	140 mm (5.5 in)	270 mm (10.6 in)	270 mm (10.6 in)
Spring free length	515.1 mm (20.3 in)	478 mm (18.8 in)	503 mm (19.8 in)
Service limit	510.0 mm (20.1 in)	473 mm (18.6 in)	468 mm (18.4 in)
Oil capacity – per leg	238 cc (8.39 fl oz)	486 cc (17.1 fl oz)	495 cc (17.51 fl oz)
Oil level	149 mm (5.87 in)	175 mm (6.89 in)	165.5 mm (6.52 in)
Fork oil grade	SAE 10W fork oil	SAE 10W fork oil	SAE 10W fork oil

Rear suspension

Wheel travel	100 mm (3.9 in)	260 mm (10.2 in)
Suspension unit:		
Travel	35 mm (1.39 in)	93 mm (3.66 in)
Spring free length	171 mm (6.73 in)	245 mm (9.65 in)
Service limit	169 mm (6.65 in)	243 mm (9.57 in)
Gas pressure	15 kg/cm² (213 psi)	15 kg/cm² (213 psi)
Swinging arm maximum free play at fork ends	1.0 mm (0.04 in)	1.0 mm (0.04 in)

Torque settings

Component	TZR model kgf m	lbf ft	DT model kgf m	lbf ft
Front fork top bolt	Not applicable		3.0	22
Front fork damper rod Allen bolt	2.4	17	6.2	45
Top yoke pinch bolts	2.0	14	2.3	17
Bottom yoke pinch bolts	3.0	22	2.3	17
Steering stem adjuster nut preload*	4.0	29	3.8	27
Steering stem adjuster nut final setting*	3.5	25	0.6	4.3
Steering stem flange nut/bolt	5.4	39	9.0	65
Clip-on handlebar to top yoke Allen bolts	0.9	6.5	Not applicable	
Clip-on handlebar pinch bolts	0.9	6.5	Not applicable	
Handlebar clamps	Not applicable		2.3	17
Swinging arm pivot shaft nut	7.4	54	9.0	65
Relay arm to frame bolt	Not applicable		5.8	42
Relay arm to connecting rod nut	Not applicable		5.8	42
Swinging arm to connecting rod nut	Not applicable		5.8	42
Rear suspension unit mounting nuts	7.4	54	3.3	24
Rear suspension unit adjuster locknut	Not applicable		5.5	40
Footrest bracket bolts	3.0	22	Not applicable	
Torque arm to swinging arm bolt	2.3	17	Not applicable	

*These settings can only be accurately obtained with the service tool specified (see Section 8). Tighten the adjuster nut to the specified preload torque, slacken it off one turn, then set to final setting.

Chapter 5 Frame and forks

1 General description

The TZR model employs a 'Deltabox' type frame, constructed of thinwall steel, and strengthened in certain areas. The DT model uses a conventional welded tubular steel frame.

Front suspension is by a pair of oil-damped, coil spring, telescopic fork legs.

Rear suspension is by slightly differing versions of the Yamaha Monocross suspension. The rear suspension unit is a De Carbon type nitrogen pressurised coil sprung unit, and the swinging arm is of box-section construction.

2 Front forks: removal

1 Support the machine securely on a strong wooden box placed underneath the engine/gearbox unit, so that the front wheel is clear of the ground. On TZR models this may require the removal of the lower fairing and exhaust system.
2 Remove the front wheel and brake caliper, as described in Chapter 6. Remove the brake hose clamp or holder and tie the caliper to the frame, to avoid straining the hydraulic hose. Place a wooden wedge between the brake pads to prevent them from being expelled if the brake lever is accidentally squeezed.
3 On TZR models it will be necessary to remove the front mudguard by removing its four mounting bolts. The combined mudguard and fork brace can then be withdrawn. The clip-on handlebars must also be removed by slackening each handlebar casting pinch bolt and removing the Allen bolts which retain the castings to the top yoke. Lift the clip-ons off the forks and support the handlebar castings to prevent straining the cables or the possible leakage of brake fluid from the master cylinder reservoir.
4 Slacken the top and bottom yoke pinch bolts and remove the forks by twisting them and pulling downwards. If the fork legs are seized in the yokes, spray the area with penetrating oil and allow time for it to work, noting that on DT models it will be necessary to slacken the gaiter upper clamp and pull the gaiter downwards for access. In severe cases the split clamps can be sprung apart with a large flat-bladed screwdriver, although extreme care must be taken not to distort or fracture the casting. On DT models, if the fork legs are to be dismantled it is preferable to slacken their top bolts whilst they are held in the yokes.

3 Front forks: dismantling and reassembly

1 Always dismantle the fork legs separately to avoid interchanging parts and thus causing an accelerated rate of wear. Store all components in separate, clearly marked, containers.

TZR model
2 Clamp the stanchion securely in a vice equipped with soft jaws, being careful not to overtighten or score the surface of the stanchion.
3 Remove the rubber plug from the top of the stanchion to reveal the circlip and top plug. Depress the top plug with a suitably sized bar and remove the circlip. Gradually release pressure on the top plug to allow it to be slowly forced out of the stanchion by the fork spring.
4 Withdraw the fork spring. Remove the fork from the vice and invert it over a drain tray. Pump the stanchion vigorously until the fork oil has drained. Remove the dust seal and using a thin-bladed screwdriver remove the oil seal retaining circlip, taking care not to damage the surface of the stanchion.
5 Once the oil has drained, slacken the Allen-headed bolt which passes up through the bottom of the lower leg and into the damper rod. It is quite likely that the damper rod will tend to rotate in the lower leg and thus impede the removal of the bolt. If this problem arises, clamp the assembly in a vice using soft jaws to hold the lower leg by the caliper mounting lugs. Obtain a length of wooden dowel about $\frac{1}{2}$ inch in diameter and form a taper on one end.
6 Pass the dowel down the stanchion and press it hard against the head of the damper rod to lock it in position whilst an assistant slackens the retaining bolt. If the dowel proves difficult to hold a self-grip wrench

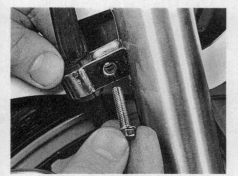

2.2 Release brake hose clamp from fork leg – TZR model shown

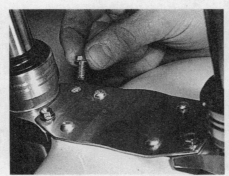

2.3a TZR125 – to permit fork removal front mudguard ...

2.3b ... and clip-on handlebars must first be removed

2.4a Slacken pinch bolts in top ...

2.4b ... and bottom yokes to release fork legs – DT model shown

2.4c DT model – slacken top bolt while fork leg is clamped in the yokes, if fork legs are to be dismantled

Chapter 5 Frame and forks

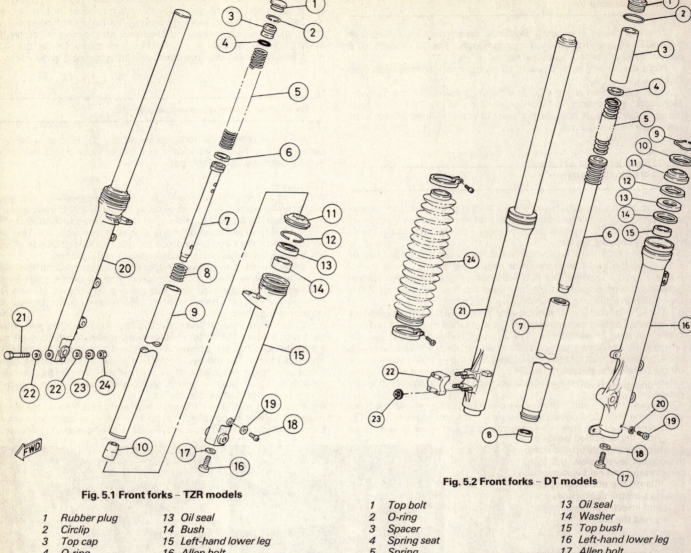

Fig. 5.1 Front forks – TZR models

1. Rubber plug
2. Circlip
3. Top cap
4. O-ring
5. Spring
6. Piston ring
7. Damper rod
8. Rebound spring
9. Stanchion
10. Damper rod seat
11. Dust seal
12. Circlip
13. Oil seal
14. Bush
15. Left-hand lower leg
16. Allen bolt
17. Sealing washer
18. Drain screw
19. Sealing washer
20. Right-hand fork leg
21. Bolt
22. Washer – 2 off
23. Spring washer
24. Nut

Fig. 5.2 Front forks – DT models

1. Top bolt
2. O-ring
3. Spacer
4. Spring seat
5. Spring
6. Damper rod and rebound spring
7. Stanchion
8. Lower bush
9. Circlip
10. Washer
11. Dust seal
12. Washer
13. Oil seal
14. Washer
15. Top bush
16. Left-hand lower leg
17. Allen bolt
18. Sealing washer
19. Drain screw*
20. Sealing washer*
21. Right-hand fork leg
22. Spindle clamp
23. Nut – 4 off
24. Gaiter

*Not fitted to 3RN8 model

or similar can be used to obtain sufficient leverage and pressure. Once the retaining bolt has been removed withdraw the stanchion assembly from the lower leg.

7 The fork oil seal can be levered out of position using a large flat-bladed screwdriver, whilst taking care not to mark the surface of the fork leg. The guide bush can also be removed in a similar way, again exercising great care. Finally, tip the damper rod seat out of the lower leg.

8 After checking the fork components for wear or damage as described in the next section, reassembly can commence. Ensure that all components are clean and that they are fitted to the fork from which they were removed. Always fit a new oil seal. The bush guide can be reused only if in perfect condition although it is recommended that it be renewed as a precautionary measure.

9 Fit the rebound spring over the damper rod and drop it into place in the stanchion. Pass a length of dowel (as described above) up through the stanchion to hold the damper rod in place, then fit the damper rod seat. Oil the stanchion and support it vertically with the damper rod uppermost. Slide the lower leg over the stanchion assembly and refit the damper rod bolt using a new sealing washer under its head. Use Loctite on the bolt threads and tighten to the specified torque setting.

10 Oil the guide bush and slide it down over the stanchion. To fit the bush in its recess it will be necessary to devise an alternative to the tubular drift tool used by Yamaha dealers. The best method is to use a length of tubing slightly bigger in diameter than the stanchion. Place a large plain washer against the bush and then tap it home using the tube as a form of slide-hammer. Take care not to scratch the stanchion during this operation; it is best to make sure that the stanchion is pushed fully inwards to that any accidental scoring is confined to the area above the seal.

11 Once the guide bush is seated, the new oil seal can be fitted. Lubricate the seal lips, then slide it down over the stanchion. Take care not to damage the seal lips during fitting. The seal should be tapped gently and squarely into the top of the lower leg, using a tubular drift. Check that the seal remains square to the lower leg and that it seats fully, then fit the retaining circlip and slide the dust seal into place.

Chapter 5 Frame and forks

3.4a TZR model – note which way up fork springs are fitted on removal

3.4b Prise out oil seal retaining circlip using a small screwdriver

3.7a Use a large flat-bladed screwdriver to lever out oil seal ...

3.7b ... and guide bush – protect lower leg as shown

3.9a Fit rebound spring to damper rod and insert into stanchion upper end

3.9b Place damper rod seat over damper rod end ...

12 Add the specified grade and amount of fork oil to the fork and slowly pump the fork up and down to distribute the oil. Check the oil is at the correct level below the top of the stanchion, with the fork fully compressed and held vertically – see Routine maintenance.
13 Clamp the stanchion securely in a vice and refit the fork spring, making sure the tighter coils are uppermost. Fit a new O-ring to the top plug and insert it in the stanchion. Depress the plug and fit the circlip. Finally refit the cap.

DT model

14 Remove the gaiter by slackening the top and bottom clamps and sliding it off the stanchion.
15 Clamp the stanchion in a vice equipped with soft jaws, being careful not to overtighten or score the surface of the stanchion. Unscrew the top bolt from the stanchion, whilst taking care not to allow it to be expelled forcibly by spring pressure, as the last threads of the bolt are unscrewed. The spacer, spring seat and fork spring can now be withdrawn from the stanchion.
16 Invert the fork leg over a suitable container and vigorously pump the leg to remove as much oil as possible.
17 Once the oil has drained, slacken the Allen-headed bolt which passes up through the bottom of the lower leg and into the damper rod. It is quite likely that the damper rod will tend to rotate in the lower leg and thus impede the removal of the bolt. If this problem arises, clamp the assembly in a vice using soft jaws to hold the lower leg by the caliper mounting lugs. Obtain a length of wooden dowel about ½ inch in diameter and form a taper on one end.
18 Pass the dowel down the stanchion and press it hard against the head of the damper rod to lock it in position whilst an assistant slackens the retaining bolt. If the dowel proves difficult to hold a self-grip wrench or similar can be used to obtain sufficient leverage and pressure. Once the retaining bolt has been removed using a thin-bladed screwdriver, remove the circlip and washer from the top of the lower leg, taking care not to damage the stanchion. Invert the fork and tip out the damper rod components.
19 To separate the stanchion from the lower leg it will be necessary to displace the top bush and the oil seal. The lower bush should not pass through the top bush, and this can be used to good effect. Push the stanchion gently inwards until it stops against the damper rod seat. Take care not to do this forcibly or the seat may be damaged. Now pull the stanchion sharply outwards until the lower bush strikes the top bush. Repeat this operation until the top bush and seal are tapped out of the lower leg. With the stanchion removed the dust seal, washer, oil seal, washer and top bush can be slid off its upper end.
20 After checking the fork components for wear or damage as described in the next section, reassembly can commence. Ensure that all components are clean and that they are fitted to the fork from which they were removed. Always fit a new oil seal. The bushes can be reused only if in perfect condition although it is recommended that they too are renewed as a precautionary measure. The lower bush is split to allow fitting over the end of the stanchion. Do not open the split any more than is essential to ease it into place.
21 Fit the rebound spring over the damper rod and drop it into place in the stanchion. Pass a length of dowel up through the stanchion to hold the damper rod in place, then fit the damper rod seat. Oil the stanchion and bush, and support it vertically with the damper rod uppermost. Slide the lower leg over the stanchion assembly and refit the damper rod bolt using a new sealing washer under its head. Using Loctite on the bolt threads and tighten to the specified torque setting.
22 Oil the top bush and slide it down over the stanchion. To fit the bush in its recess it will be necessary to devise an alternative to the tubular drift tool used by Yamaha dealers. The best method is to use a length of tubing slightly bigger in diameter than the stanchion. Place a large plain washer against the bush and then tap it home using the tube as a form of slide-hammer. Take care not to scratch the stanchion during this operation; it is best to make sure that the stanchion is pushed fully inwards so that any accidental scoring is confined to the area above the seal.
23 Once the top bush is seated, the new oil seal can be fitted. Fit the plain washer with the chamfered edge facing upwards. Lubricate the seal lips, then slide it down over the stanchion. Take care not to damage the seal lips during fitting. The seal should be tapped gently and squarely into the top of the lower leg, using a tubular drift. Check that the seal remains square to the lower leg and that it seats fully, then fit the washer and slide the dust seal into place, followed by another washer and circlip.

Chapter 5 Frame and forks

3.9c ... and insert stanchion assembly into lower leg

3.9d Apply thread-locking compound to damper rod Allen bolt and tighten to specified torque setting

3.11 Use a tubular drift to tap guide bush and a new seal into place

3.13 Ensure circlip is correctly located in its groove

3.19 DT model – oil seal, washers, dust seal and top bush should be removed with stanchion

3.20 Bottom bush is split for easy removal

3.22 Oil top bush liberally before sliding into place

3.25a Refit fork spring and spring seat ...

3.25b ... followed by spacer

24 Add the specified grade and amount of fork oil to the fork and slowly pump the fork up and down to distribute the oil. Check the oil is at the correct level below the top of the stanchion, with the fork fully compressed and held vertically – see Routine maintenance.

25 Clamp the stanchion securely in a vice and refit the fork spring, spring seat, spacer and top bolt. Note that the top bolt can be tightened to the specified torque at this stage if it can be held firmly enough, but do not risk overtightening the stanchion. A better method is to tighten the top cap when the stanchions have been refitted to the machine and are securely clamped in the yokes.

All models

26 When reassembly is complete refit the fork legs into the fork yokes as described in Section 5.

4 Front forks: examination and renovation

1 If the forks have been damaged in an accident, it is essential to inspect both fork yokes, the stanchions and the lower legs, for distortion and hairline cracks. Distorted components must be renewed, do not attempt to straighten them.

2 The parts most likely to wear are the sliding surfaces of the bush(es). These control the play in the forks and are designed to wear before damage occurs to the stanchion or lower leg. If there are signs of scoring or obvious wear, the bush(es) must be renewed. Only in extreme cases will the stanchion or lower leg be worn; in these cases the affected item must be renewed.

3 Check the stanchion for signs of scoring. Damage of this type can be

Chapter 5 Frame and forks

caused by dirt trapped below a damaged or worn dust seal and can be avoided by ensuring that it is renewed whenever the oil seal is renewed. If there has been impact damage, check that the stanchions are straight by rolling them on a flat surface. A bent stanchion must be renewed; do not attempt to straighten it.

4 The oil seals should be renewed whenever they are disturbed, as should all sealing O-rings and sealing washers. Check carefully the condition of each damper rod piston ring and renew it if there is any doubt about its condition. Where fitted, check the dust excluder and fork gaiters for signs of wear or damage and renew them if necessary.

5 Measure the spring free lengths; if either has settled to less than the specified length both springs must be renewed.

6 Thoroughly clean all components and dry them ready for reassembly.

5 Front forks: refitting

1 Refitting the forks is accomplished by reversing the removal sequence. Slide the forks back into their correct position in the yokes and temporarily tighten the pinch bolts.

TZR model

2 Refit the handlebar castings, and tighten the casting to top yoke Allen bolts to the specified torque setting. Position the stanchions so that they protrude 10 mm (0.4 in) above the top of the handlebar castings and tighten the handlebar pinch bolts to the specified torque setting.

3 Tighten the top and bottom yoke pinch bolts to their specified torque settings. Refit the mudguard.

DT model

4 Position each stanchion so that its top edge is flush with the top surface of the top yoke, and tighten the bottom yoke pinch bolts to the specified torque setting.

5 If the fork has been dismantled, the top bolt should now be tightened to the specified torque, followed by the top yoke pinch bolt. Repeat the sequence on the other fork leg.

All models

6 Refit the front wheel and brake caliper as described in Chapter 6. Refit the brake hose clamp or holder and move the machine off its support. Thoroughly check the operation of the front forks and brake before taking the machine out on the road.

6 Steering head: dismantling

1 Before dismantling the steering head, it is advisable to remove the fuel tank to prevent damage to its paintwork.

2 Remove the front forks as described in Section 2 of this chapter.

TZR model

3 If a fairing is fitted this must first be removed as described later in this Chapter. Remove the headlamp unit from its shell by releasing the two retaining screws, unplug the headlamp and parking lamp wiring connectors and lift the unit clear.

4 Disconnect all the electrical leads from inside the headlamp shell and remove the shell from its mounting brackets. Release the knurled rings retaining the speedometer and tachometer cables to the underside of the instrument panel and pull the cables free.

5 Remove its retaining bolt and disconnect its wiring to remove the horn. Remove the front brake hose clamp from the headlamp stay, then release the two headlamp stay mounting bolts and manoeuvre the stay clear of the yokes.

6 Remove the steering stem top bolt. The top yoke and instrument

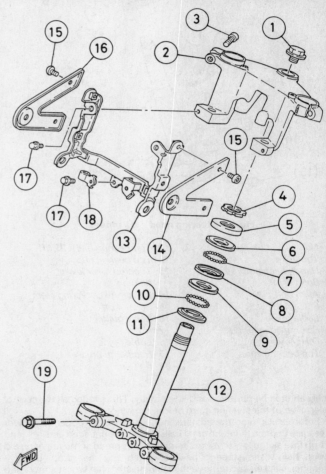

Fig. 5.3 Steering head – TZR models

1 Top bolt
2 Top yoke
3 Allen bolt – 2 off
4 Bearing adjuster nut
5 Dust cover
6 Top cone
7 Top bearing ball – 19 off
8 Top bearing cup
9 Lower bearing cup
10 Lower bearing ball – 19 off
11 Lower bearing cone
12 Lower yoke/steering stem
13 Headlamp stay
14 Left-hand headlamp bracket
15 Screw – 4 off
16 Right-hand headlamp bracket
17 Bolt – 3 off
18 Clamp
19 Bolt – 2 off

5.2 TZR model – stanchions should protrude 10 mm above handlebar castings

area and remove all cable ties and bands. Unscrew their knurled rings and pull the speedometer and tachometer cables from the underside of the instrument panel. The panel can now be removed by releasing the two bolts which retain the instruments to the top yoke.

12 Remove the four handlebar clamp bolts, and remove the clamps. The handlebars can now be released and moved rearwards to rest on the frame top tubes. Support the handlebars to prevent any strain being placed on the control cables or hydraulic hose.

13 Remove the three headlamp stay mounting bolts and manoeuvre the stay clear of the yokes.

14 Prise the rubber cap off the steering stem flange nut and remove the nut, followed by the steering stem pinch bolt. The top yoke, complete with ignition switch can now be removed. Try to lodge all the control cables clear of the steering stem before further dismantling.

15 Using a C-spanner, slacken and remove the adjuster nut whilst supporting the bottom yoke. Lift away the nut, dust cover and top cone, then remove the steel balls from the upper race and place them in a container for safekeeping.

16 Lower the bottom yoke and steering stem out of the headstock. The lower bearing is a taper roller bearing which will remain on the steering stem.

7 Steering head bearings: removal, examination and refitting

1 For straight line steering to be consistently good, the steering head bearings must be absolutely perfect. Even the smallest amount of wear may cause steering wobble at high speeds and judder during front wheel braking.

Cup and cone bearings

2 The ball bearing tracks of the respective cup and cone bearings should be polished and free from indentations, cracks or pitting. If signs of wear are evident, the cups and cones must be renewed. The cups are an interference fit in the steering head and can be tapped from position with a suitable drift. Tap firmly and evenly around each cup to ensure that it drives out squarely. It may prove advantageous to curve the end of the drift slightly to improve access. Note that with this method there is a real risk of damage unless care is taken. If the race refuses to move, **stop**. Leave the job until a proper bearing extractor can be obtained. The new cups can be pressed into the head using the drawbolt arrangement shown in the accompanying figure, or by using a large diameter drift which bears only on the outer edge of the cups.

3 The lower bearing cone on the TZR model is a drift fit on the steering stem and will require levering off the stem. Use two screwdrivers placed on opposite sides of the cone to work it free. A length of tubing, with an internal diameter slightly larger than the steering stem may be

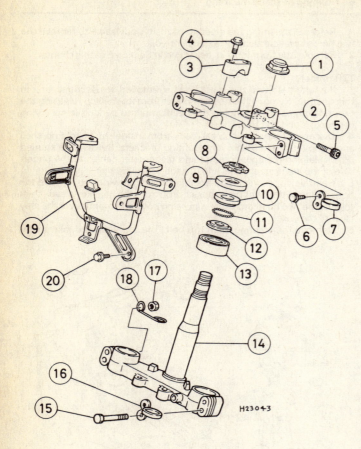

Fig. 5.4 Steering head – DT models

1 Nut	11 Top bearing ball – 22 off
2 Top yoke	12 Top bearing cup
3 Handlebar clamp – 2 off	13 Tapered roller lower bearing
4 Bolt – 4 off	
5 Allen bolt	14 Lower yoke/steering stem
6 Bolt – 4 off	15 Bolt – 4 off
7 Cable guide	16 Cable guide
8 Bearing adjuster nut	17 Nut
9 Dust cover	18 Cable guide
10 Top bearing cone	19 Headlamp bracket
	20 Bolt

panel can then be removed and lifted away. Try to lodge all the control cables clear of the steering stem before removing it.

7 Before slackening the adjuster nut note that the upper and lower races each contain nineteen steel balls. These are uncaged and will tend to drop free as the lower yoke is removed, so spread some rag or an old blanket below the steering head to catch any that drop. Using a C-spanner, slacken and remove the steering stem nut whilst supporting the bottom yoke. Lift away the nut, dust cover and the top cone, then remove the steel balls from the upper race.

8 Carefully lower the yoke and steering stem, trying not to dislodge the balls in the lower race. Remove the balls from the lower race and place them in a container for safekeeping.

DT model

9 Release the four front mudguard mounting bolts and washers and remove the mudguard. The headlamp fairing should also be removed; this is retained by two bolts, one each side of the fairing.

10 Remove the headlamp unit from its mountings after releasing the two retaining bolts. Unplug the headlamp and parking lamp wiring connectors and lift the unit clear.

11 Disconnect all the electrical leads around the headlamp bracket

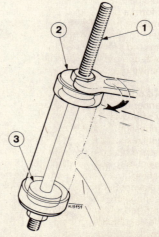

Fig. 5.5 Fabricated drawbolt tool for fitting steering head bearing outer races

1 Drawbolt
2 Thick washer
3 Guide

used as a drift to tap the new bottom cone firmly onto its seat, make sure that the drift does not damage the bearing surface in any way.
4 Ball bearings are relatively cheap. If the originals are marked or discoloured they **must** be renewed. To hold the steel balls in place during reassembly of the steering head, pack the bearings with grease. Although a small gap will remain when the balls have been fitted, on no account must an extra ball be inserted, as the gap is intended to prevent the balls from skidding against each other and wearing quickly.
5 On reassembly, pack grease around the bottom cone and stick the balls to it as described above. Grease the inside of the headstock and the steering stem to prevent corrosion. Pack grease into the top cup, refit the balls and place the top cone in position. Remember to hold the cone steady while the bottom yoke is refitted. Refit the steering stem adjuster nut and tighten it by hand only at this stage.

Taper roller bearings
6 The lower steering head bearing on the DT model is of the taper roller type and should be examined using the following procedure.
7 The inner race is easily checked after all traces of old grease have been removed by washing in a suitable solvent. Turn the race slowly, checking for marks or discoloration of the roller faces.
8 Clean the outer race, and examine the bearing surface for wear or damage. If either are worn, the bearing must be renewed. The outer race is a fairly tight fit in the headstock. Most universal slide-hammer type bearing extractors will work here, and these can often be hired from tool shops. Alternatively, a long drift can be used to drive out the outer race; see cup and cone bearing text.
9 The lower inner race can be levered off the steering stem, using screwdrivers on opposite sides to work it free. The dust seal should now be removed and examined. If the seal is worn or damaged it must be renewed.
10 To fit the new bearing, find a length of tubing slightly larger in its internal diameter than the steering stem. This will suffice as a tubular drift. Grease the bearing thoroughly and wipe a trace of grease around the steering stem. Drive the bearing home evenly and fully.
11 The new outer race can be installed using a home-made version of the drawbolt arrangement shown in Fig. 5.5.

8 Steering head: reassembly

1 The steering head should be reassembled in the reverse order of the dismantling sequence, noting the following points:
2 To preload the bearings to the torque setting specified by the manufacturer (see Specifications) it will be necessary to use the service tool, Pt. No. 90890-01403, which consists of a short C-spanner to which a torque wrench should be attached. Position the C-spanner and torque wrench so that they form a right angle. After preloading the bearings, slacken the adjuster nut one full turn and then tighten to the final setting. Note that it is important to check the feel of the steering afterwards, as described below; if it is too tight readjust the bearings as follows. Note that because the tool is of a pre-determined length it is not possible to substitute any alternative means of tightening to the specified torque setting.
3 If the service tool is not available, tighten the adjuster nut hard using a conventional C-spanner to preload the bearings. In each case slacken the nut immediately afterwards and carry out bearing adjustment as follows: Slacken the nut slightly until pressure is just released, then turn it slowly clockwise until resistance is **just** evident. Take great care not to apply excessive pressure because this will cause premature failure of the bearings. The object is to set the adjuster so that the bearings are under a **very light** loading, just enough to remove any free play.
4 Referring to the relevant Section of Routine maintenance check that the steering head bearings are correctly adjusted as soon as the forks and front wheel are refitted, but before the handlebars are refitted.
5 When refitting the handlebars on DT models, the handlebar clamps should be installed with the punch mark facing forward. Tighten the front clamp bolts first then the rear bolts, the gap at the rear of the handlebar clamp is intentional.
6 Tighten all bolts to the specified torque settings given at the start of this Chapter.
7 Check the operation of the front forks, brake, electrical system and controls before taking the machine out on the road.

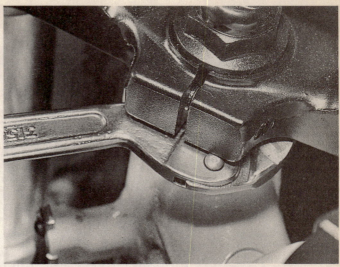
8.3 Use a C-spanner to adjust steering head bearings

9 Frame: examination and renovation

1 The frame is unlikely to require attention unless accident damage has occurred. In some cases, renewal of the frame is the only satisfactory course of action if it is badly out of alignment. Only a few frame repair specialists have the jigs and mandrels necessary for resetting the frame to the required standards of accuracy and even then there is no easy means of assessing to what extent the frame may have been overstressed.
2 After the machine has covered a considerable mileage, it is advisable to examine the frame closely for signs of cracking or splitting at the welded joints. Rust can also cause weakness at these joints. Minor damage can be repaired by welding or brazing, depending on the extent and nature of the damage.
3 Remember that a frame which is out of alignment will cause handling problems and may even promote 'speed wobbles'. If misalignment is suspected, as the result of an accident, it will be necessary to strip the machine completely so that the frame can be checked and, if necessary, renewed.

10 Swinging arm and suspension linkage: removal

1 Referring to Chapter 5, disconnect the final drive chain and remove the rear wheel. Remove the rear suspension unit as described in Section 13 of this Chapter.

TZR model
2 Remove the brake hose guide from the swinging arm on 3PC2 models and support the caliper securely. Remove the nut and washer from the swinging arm pivot shaft and withdraw the shaft from the opposite side. The swinging arm can now be removed from the frame. If the shaft has stuck firmly in place with corrosion apply a penetrating fluid, such as WD40, and allow time for this to work. Rotate the pivot shaft head in an attempt to free it, or in stubborn cases use a long drift to drive out the pivot shaft.

DT model
3 Unbolt the rear mudguard from the frame and remove all brake hose guides from the right-hand side of the swinging arm. Remove the two bolts which hold the brake caliper protector and remove it together with the caliper assembly. Support the caliper carefully so that no undue strain is placed on the hydraulic hose. It is good practice to insert a wooden wedge or similar component between the brake pads to prevent their accidental ejection.
4 Remove both the chainguard, from the top of the swinging arm, and the chain guide from the underside of the swinging arm.
5 Remove the nut and washer from the swinging arm pivot shaft and withdraw the shaft from the opposite side. The swinging arm can now

106 Chapter 5 Frame and forks

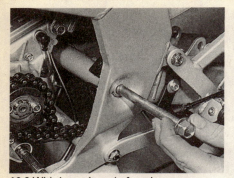
10.2 Withdraw pivot shaft and remove swinging arm – TZR model shown

10.6a To release relay arm, remove bolt which secures connecting arms to relay arm ...

10.6b ... and relay arm to frame mounting bolt

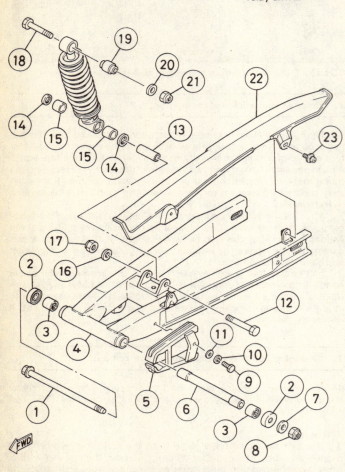

Fig. 5.6 Swinging arm and rear suspension unit – TZR models

1	Pivot shaft	12	Bolt
2	Grease seal – 2 off	13	Inner sleeve
3	Bearing – 2 off	14	Grease seal – 2 off
4	Swinging arm	15	Bush – 2 off
5	Chain buffer	16	Washer
6	Sleeve	17	Nut
7	Washer	18	Bolt
8	Nut	19	Bush
9	Bolt	20	Washer
10	Spring washer	21	Nut
11	Washer		

be removed from the frame. If the pivot shaft has stuck firmly in place release it as described above for the TZR model.
6 Remove the bolt which secures the two connecting arms to the relay arm and remove both arms. Then remove the pivot bolt which mounts the relay arm to the frame and remove the relay arm from the frame.

11 Swinging arm and suspension linkage: examination and renovation

1 Thoroughly clean all components, removing all traces of dirt, corrosion and grease.
2 Inspect all components closely, looking for obvious signs of wear such as heavy scoring, or for damage such as cracks or distortion due to accidental impact. Any obviously damaged or worn component must be renewed. If the painted finish has deteriorated it is worth taking the opportunity to repaint the affected area, ensuring that the surface is correctly prepared beforehand.
3 Check the pivot shaft for wear. If the shank of the shaft is seen to be stepped or badly scored, it must be renewed. Remove all traces of corrosion and hardened grease from the shaft before checking it for straightness by rolling it on a flat surface, such as a sheet of plate glass, if it is not perfectly straight it must be renewed. Check also that its threads and those of the retaining nut are in good condition.
4 Lever out the grease seals, using a flat-bladed screwdriver, and examine them for signs of wear or damage and renew if necessary. Worn bushes or bearings in either the swinging arm or the various linkage pivots can be drifted out of their bores, but note that removal will destroy them; new bushes should be obtained before work commences. The new bushes should be pressed or drawn into their bores, rather than driven into place. In the absence of a press, a suitable drawbolt arrangement can be made up as described below.
5 It will be necessary to obtain a long bolt or a length of threaded rod from a local engineering works or some other supplier. The bolt or rod should be about 1 inch longer than the combined length of the cross tube and one bush. Also required are suitable nuts and two large and robust washers. In the case of threaded rod, fit one nut to one end of the rod and if required, stake it in place for convenience.
6 Fit one of the washers over the bolt or rod so that it rests against the head, then pass the assembly through the cross-tube. Over the projecting end place the bush, which should be greased to ease installation, followed by the remaining washer and nut. Holding the bush to ensure that it is kept square, slowly tighten the nut so that the bush is drawn into the cross-tube. Once it is fully home, remove the drawbolt arrangement and repeat the sequence to fit the remaining bush.
7 Inspect the chain buffer fitted to the left-hand lug of the swinging arm. If this shows any sign of wear it should be renewed.

DT models only
8 On DT models it will be necessary to check the swinging arm side clearance before reassembly. Remove the inner sleeves from the swinging arm lugs by gently tapping them out. Assuming that the sleeves are otherwise undamaged, measure their length. In both cases this should be within the range of 66.45 – 66.75 mm (2.616 – 2.628 in); if not, the sleeves must be renewed along with their bushes. Measure the width of the rear engine mounting boss, which the swinging arm pivot shaft passes through, and add it to the length of both inner sleeves.
9 Check the thickness of the plain washers which fit inside the thrust covers. These should be within the range of 0.7 – 0.9 mm (0.028 – 0.035

Chapter 5 Frame and forks

Fig. 5.7 Swinging arm, suspension linkage and unit – DT models

1. Swinging arm
2. Pivot shaft
3. Thrust cover – 2 off
4. Shim – as required
5. Washer – 2 off
6. Grease seal – 2 off
7. Inner sleeve – 2 off
8. Bearing – 2 off
9. Bush – 2 off
10. Grease seal – 2 off
11. Grease nipple
12. Washer
13. Nut
14. Chain guide
15. Bolt – 2 off
16. Chain buffer
17. Bolt – 2 off
18. Washer – 2 off
19. Chainguard
20. Bolt – 2 off
21. Screw – 2 off
22. Washer – 2 off
23. Relay arm
24. Grease nipple
25. Bolt
26. Grease seal – 2 off
27. Bearing – 2 off
28. Inner sleeve
29. Washer
30. Bolt
31. Grease nipple – 2 off
32. Pivot bolt – 2 off
33. Inner sleeve – 2 off
34. Grease seal – 4 off
35. Bush – 4 off
36. Right-hand connecting arm
37. Washer – 2 off
38. Nut – 2 off
39. Left-hand connecting arm
40. Suspension unit
41. Bolt
42. Headed insert – 2 off
43. Washer
44. Nut
45. Bolt
46. Bush
47. Washer
48. Nut

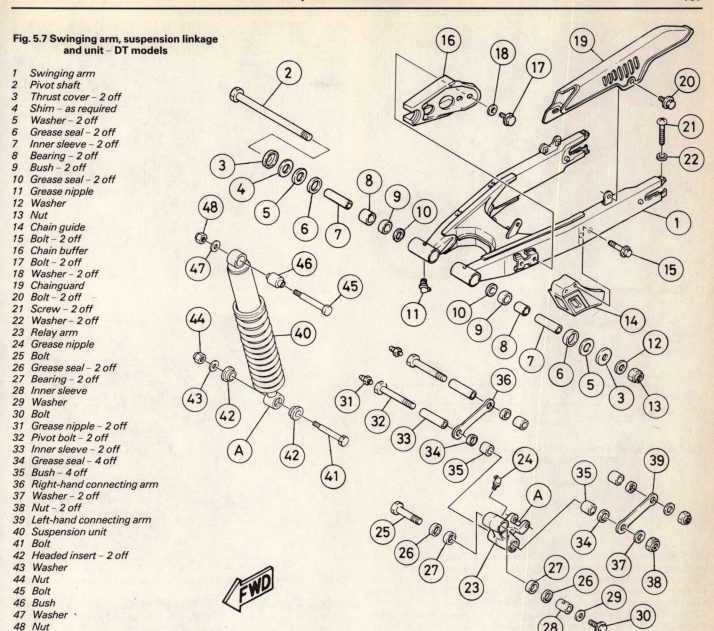

11.2 Examine all components for signs of wear or damage

11.4 Bushes and bearings can be pressed out of position if renewal is required

11.8 Measuring width of the rear engine boss

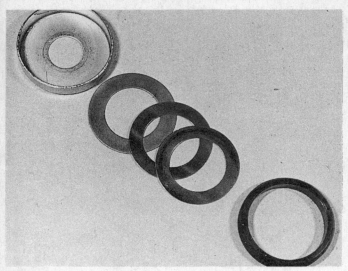

11.9a Side clearance is adjusted using shims ...

11.9b ... which fit inside thrust caps

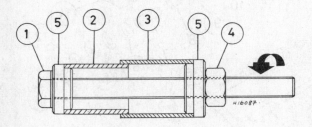

Fig. 5.8 Fabricated drawbolt tool for removing or refitting swinging arm and rear suspension bushes

1 Drawbolt 4 Nut
2 Bush 5 Washers
3 Housing

12.1a Lubricate all bushes, bearings and sleeves with specified grease

12.1b Grease inside of thrust caps before refitting them to the swinging arm

in), if not they must be renewed. Measure the pivot width of the swinging arm and add it to the thickness of both washers. Subtract this figure (two plain washers + swinging arm pivot) from the one obtained earlier (two inner sleeves + engine boss) to obtain the swinging arm side clearance. The side clearance should be between 0.4 – 0.7 mm (0.016 – 0.028 in). If not, this can be adjusted by fitting shims between the thrust caps and plain washers. If required, shims can be purchased from an authorized Yamaha dealer. When fitting the shims, they should be positioned evenly on each side of the pivot. If there is an odd number of shims, put the extra one on the right-hand side of the swinging arm.

12 Swinging arm and suspension linkage: refitting

1 The swinging arm is refitted by reversing the removal sequence. Make sure all bearings, bushes and seals are refitted in their correct positions, referring to the accompanying illustrations and photographs for guidance. Lubricate all seals, bearings, bushes, inner sleeves and pivot bolts with a lithium-based grease. In the case of DT models ensure that any shims necessary to obtain the correct side clearance have been fitted (see previous section). Apply a small amount of grease to the inside of the thrust caps before fitting the caps to the swinging arm.
2 Refit the relay arm and connecting rods (DT models only). Install the swinging arm in the frame and refit the pivot shaft, followed by the washer and nut. Tighten the nut to the specified torque setting and check that the swinging arm moves up and down freely and does not have an excessive amount of side play.
3 Reconnect the rear suspension unit and refit the wheel, referring to the relevant sections of Chapters 5 and 6 for further information.

Chapter 5 Frame and forks

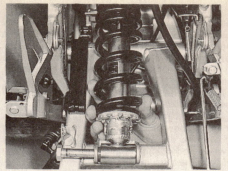

13.3a TZR model – withdraw suspension unit lower mounting bolt ...

13.3b ... followed by upper mounting bolt and remove unit

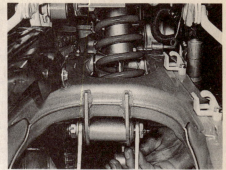

13.7 DT model – lower suspension unit out through space between swinging arm and relay arm

Tighten all bolts to their specified torque settings. On DT models, lubricate the swinging and relay arms by pumping grease into the grease nipples provided.
4 Check that the rear suspension moves smoothly before taking the machine out on the road.

13 Rear suspension unit: removal

1 Remove the dualseat and both sidepanels.

TZR model
2 Remove the rear wheel, as described in Section 4 of Chapter 6. Note that it is necessary to remove the left-hand footrest mounting bracket and chainguard to gain access to the lower suspension unit mounting bolt. First remove the gearchange lever pivot bolt from the footrest bracket, then remove both bracket mounting bolts and lift it clear of the frame. Remove the two chainguard retaining screws and remove the guard.
3 Slacken and remove the lower suspension unit mounting bolt, followed by the upper one. The suspension unit can now be removed from the frame.

DT model
4 Lift the rear wheel clear of the ground by placing a suitable support under the engine/gearbox unit. Remove the rear wheel as described in Chapter 6. Remove fuel tank as described in Chapter 3.
5 Remove the bolt which secures the two connecting arms to the underside of the swinging arm and rotate the arms clear of the lower suspension unit mounting bolt.
6 Slacken and remove the lower suspension unit mounting bolt whilst holding the swinging arm, so that it does not drop downwards once the bolt is removed. Then remove the upper suspension unit mounting bolt.
7 The rear suspension unit can now be removed by lifting the swinging arm and lowering the unit out through the space between the swinging arm and relay arm.

14 Rear suspension unit: examination and renovation

1 Examine the rear suspension unit for signs of oil leakage or damage. The unit is of sealed construction and cannot be repaired; if faulty the unit must be renewed.

2 Should it become necessary to dispose of the cylinder do not just throw it away. It is first necessary to release the gas pressure and the manufacturers recommend that the following procedure is followed.
3 Refer to the accompanying figure and mark a point 15 – 20 mm above the bottom of the cylinder. Place the unit securely in a vice. Wearing proper eye protection against escaping gas and/or metal particles, drill a 2 – 3 mm hole through the previously marked point on the cylinder.
4 Check all the bushes and pivot bolts for damage and corrosion and renew if necessary. If renewal is required, the bushes should be pressed out of position as described in Section 11. Oil and dust seals must also be examined for damage or deterioration; renew if necessary.

15 Rear suspension unit: refitting and adjustment

1 The rear suspension unit is refitted by reversing the removal sequence. Apply a lithium-based grease to all oil seals, bushes and to the inside surface of all washers. Tighten all bolts to their specified torque settings given in the specifications at the start of this Chapter.
2 Thoroughly check the operation of the rear suspension before taking the machine on the road.
3 The suspension unit spring preload can be adjusted to suit the rider's weight, riding style and road condition.
4 On the TZR model, adjustment is made by turning the cam on the lower part of the unit. There are six positions and adjustment can be made by turning the cam to the right to increase preload and harden the

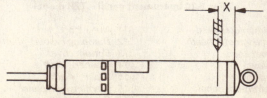

Fig. 5.9 Position of drilling on rear suspension unit

Dimension x – 15 to 20 mm

14.4 Examine suspension unit bushes for wear or corrosion – renew if necessary

Chapter 5 Frame and forks

ride, or to the left to decrease preload and soften the ride, viewed from the right-hand side of the machine. Make sure the cam recess locates fully after adjustment. Note that the standard setting is one position down from the softest position.

5 On the DT model, spring preload is measured in terms of the spring's installed length. The standard setting is 230 mm (9.1 in), although this can be adjusted to a minimum of 220 mm (8.7 in) and a maximum of 235 mm (9.3 in). Adjustment is made by slackening the locknut at the base of the unit and turning the adjuster nut above. When viewed from the right-hand side of the machine, turning the nut to the right increases preload and hardens the ride, whereas turning it to the left decreases preload and softens the ride. Note that one full turn of the nut equates to 1 mm of spring preload. Check that the spring length does not exceed either the minimum or maximum setting before tightening the locknut to the specified torque setting.

16 Footrest, stands and controls: examination and renovation

1 At regular intervals all footrests, the stand, the brake pedal and the gear lever pivots should be checked and lubricated. Check that all mounting nuts and bolts are securely fastened, using the recommended torque wrench settings where these are given. Check that any securing split pins are correctly fitted.

2 Check that the bearing surfaces at all pivot points are well greased and unworn, renewing any component that is excessively worn. If lubrication is required, dismantle the assembly to ensure that grease can be packed fully into the bearing surface. Return springs, where fitted, must be in good condition with no traces of fatigue and must be securely mounted.

3 If accident damage is to be repaired, check that the damaged component is not cracked or broken. Such damage may be repaired by welding, if the pieces are taken to an expert, but since this will destroy the finish, renewal is usually the most satisfactory course of action. If a component is merely bent it can be straightened after the affected area has been heated to a full cherry red, using a blowlamp or welding torch.

17 Instrument panel: removal and refitting

TZR model

1 If a fairing is fitted this must first be removed as described in Section 20. Remove the headlamp unit from its shell by releasing the two retaining screws, unplug the headlamp and parking lamp wiring connectors and lift the unit clear. Remove the bolts which mount the shell to its bracket and position it clear of the instrument panel.

2 Remove the four R-clips and washers and lower the bottom cover away from the instruments; this gives access to the panel bulbs which can be renewed if necessary. Disconnect the instrument panel block connector from the main wiring inside the headlamp shell and release the knurled rings which retain the drive cables. The complete instrument panel can now be lifted clear of its mountings and the speedometer and tachometer heads removed.

DT model

3 To gain access to the instrument panel it is first necessary to remove the headlamp fairing. This is retained by two screws, one each side of the fairing.

4 Remove the bolts which mount the headlamp unit to its bracket, unplug the headlamp and parking lamp wiring connectors and lift the unit clear.

5 Disconnect all the electrical wiring connectors between the instrument panel and main wiring loom and release the speedometer and tachometer cables from the underside of the instruments. The panel can now be removed by releasing the two bolts which mount the instruments to the top yoke.

6 Remove the four R-clips and washers from the underside of the panel. The speedometer and tachometer can now be pulled out of their mountings and access gained to the panel bulbs. If the instruments are to be removed completely, unplug all the bulbs, noting where each one is fitted. It will also be necessary to disconnect the three temperature gauge wires if the tachometer is to be removed, again note which wire goes to each terminal so that it can be correctly rewired on reassembly.

All models

7 The instrument head itself is generally reliable, and is the least likely culprit in the event of failure, this normally being attributed to the cable rather than the instrument mechanism. If however, it is noted that the speedometer has ceased to function whilst the odometer (mileage recorder) still functions, the instrument can be assumed to have failed. No form of repair is practicable at home, and a replacement speedometer will be required. The only alternative is to seek the assistance of one of the companies who specialise in this type of repair.

8 Refitting is a straightforward reversal of the removal sequence. Check that all the mounting rubbers are in a good condition, renewing them if necessary, and all bulbs and wiring connectors are correctly refitted. Spin the front wheel and kick the engine over to help engage the drive cables correctly into the instrument heads.

9 Check that the instrument panel warning lights function correctly before taking the machine out on the road.

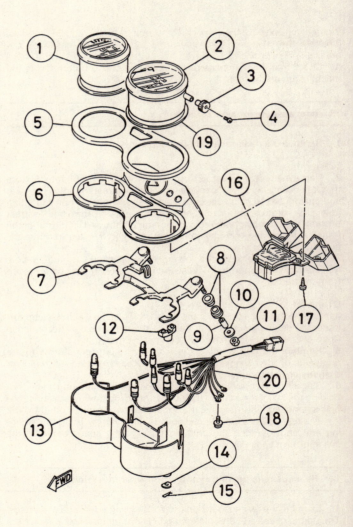

Fig. 5.10 Instrument panel – TZR model

1 Tachometer head
2 Speedometer head
3 Reset knob
4 Screw
5 Top cover
6 Instrument panel
7 Mounting bracket
8 Damping rubber – 4 off
9 Spacer – 2 off
10 Washer – 2 off
11 Nut – 2 off
12 Damping rubber – 4 off
13 Bottom cover
14 Washer – 4 off
15 R-pin – 4 off
16 Warning light panel
17 Screw – 4 off
18 Screw and washer – 3 off
19 Damping ring – 2 off
20 Instrument wiring

Chapter 5 Frame and forks

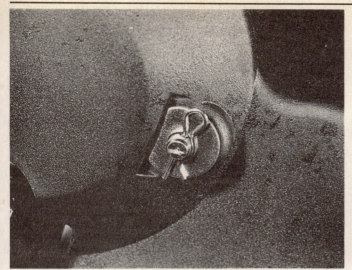
17.2a TZR model – remove four R-clips and washers and remove bottom cover

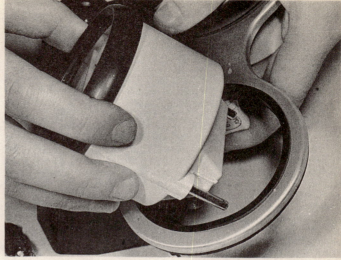
17.2d Instrument heads can then be removed from panel

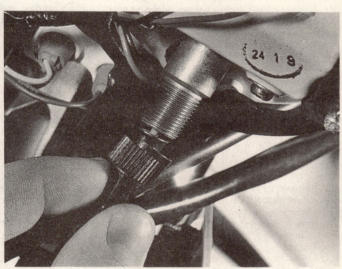
17.2b Release drive cables ...

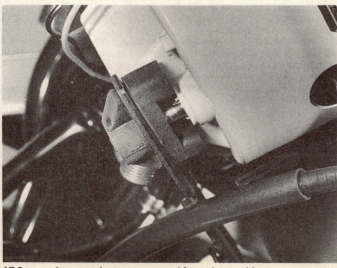
17.2c ... and remove instrument panel from the machine

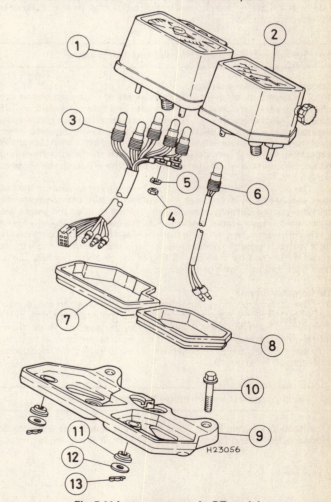

Fig. 5.11 Instrument panel – DT model

1 Tachometer head
2 Speedometer head
3 Tachometer bulb assembly
4 Nut – 3 off
5 Spring washer – 3 off
6 Speedometer bulb
7 Damping rubber
8 Damping rubber
9 Bottom cover
10 Bolt – 2 off
11 Damping rubber – 4 off
12 Washer – 4 off
13 R-pin – 4 off

18 Instrument drive cables: examination and maintenance

1 If an instrument fails to operate suddenly, or if its movement is jerky or sluggish, check that the cable is not broken, and remove the inner cable to check that it is adequately lubricated and not worn due to a trapped or kinked outer cable.
2 The cables should be removed at regular intervals so that they can be checked for wear or damage and so that the inner cables can be lubricated.
3 The cables are secured at their ends by knurled sleeve nuts which should be slackened and tightened with a pair of pliers on removal and refitting. Spin the inner cable to check for resistance. Most cables have a tight spot, but if the resistance is severe and a wavering instrument needle has been noted, the cable should be renewed. Lubrication is difficult with this type of cable, but an aerosol chain grease or a silicone-based lubricant can often be introduced using the aerosol's thin extension nozzle. Do not apply lubricant to the upper six inches of the cable otherwise there is a risk of it working up into the instrument head. On refitting, spin the front wheel (or turn the engine over on the kickstart, as applicable) to help the drive mechanism engage the inner cable.
4 Check that the cables are routed correctly with no kinks or sharp bends, and that they are secured by any guide, clamps or ties provided for this purpose.

19 Instrument drives: examination and maintenance

Speedometer drive gearbox
1 The speedometer drive gearbox is mounted on the right-hand side of the hub on DT models, and on the left-hand side of the hub on TZR models. The gearbox must be regarded as a sealed unit and requires no maintenance except for packing with the recommended grease whenever the wheel bearings are lubricated. Individual parts are not available separately on the DT model, although on TZR models the drive ring, retainer and oil seal which fit inside the front wheel hub can be obtained separately.
2 The drive ring fitted in the hub can be removed as soon as the hub dust seal has been levered out. On refitting, ensure that the drive ring tabs are located correctly in the slots in the hub and that the grease seal is renewed. Smear a small amount of grease over the drive ring and seal lips before refitting the wheel to the machine. Note that the commonest fault with this type of speedometer drive is that the drive ring's raised ears are flattened by careless refitting of the gearbox; the ears can be bent back into position if care is exercised.

Tachometer drive
3 The tachometer drive assembly consists of a worm mechanism driven by the clutch outer drum, and a gear mounted vertically in the crankcase. This mechanism is lubricated by the gearbox oil and therefore requires no maintenance at all. If the drive assembly is at fault, the right-hand crankcase cover must be removed to inspect the mechanism as described in Chapter 1.

20 Fairing: removal and refitting – TZR model

1 Although a fairing is not fitted as original equipment, Yamaha do produce a fairing and fixing kit for the TZR125. If fitted, the fairing will severely restrict access to the carburettor and engine/gearbox unit and it will be necessary to partially or totally remove it before certain operations are undertaken. The fairing can be removed using the following procedure.
2 Slacken and remove the two Allen bolts which secure the right and left-hand fairing lowers together, and remove the three Allen bolts which join the right-hand lower to the upper fairing. The right-hand lower can be removed by releasing its two mounting bolts and pulling it backwards to disengage the pins on its bottom edge.
3 Remove the three Allen bolts which join the left-hand lower and upper fairing. The left-hand lower can then be removed by releasing the two Allen bolts which secure it to its mounting brackets and lifting it clear. With both the right and left-hand fairing lowers removed full access to the engine/gearbox unit is regained.
4 The upper fairing is retained by six bolts. Before starting to remove the upper fairing it will first be necessary to remove the front turn signals. Disconnect the turn signal wires from the main wiring loom and release the nut which secures them to the fairing mounting bracket. The turn signal lamps can then be removed from the machine.
5 Remove both the fairing-mounted mirrors and then release the two bolts situated on the underside of the upper fairing, which can then be lifted clear of its mounting bracket.
6 The fairing is refitted by a reversal of the removal sequence. Check the condition of all mounting rubbers and grommets and renew them if necessary. Ensure that the mountings are correctly assembled before refitting the Allen bolts; tighten the bolts securely but avoid overtightening, otherwise the fairing may be damaged. Make sure that the pins on the bottom of the fairing lowers engage correctly when refitting the right-hand fairing lower. Finally, check the operation of the turn signals before taking the machine on the road.

21 Seat and sidepanels: removal and refitting

Seat – TZR models
1 Unlock the seat via the lock set in its left side and lift it off the motorcycle. On refitting, engage the prong on the front underside of the seat with the frame and push down on the rear of the seat to lock it in position.

Seat – DT models
2 Remove both sidepanels (see below) to access the seat mounting bolt on each side. Remove the mounting bolts, lift the seat up at the rear and disengage its front mounting prongs.
3 Refit in a reverse of the removal procedure, noting that the seat front mounting prongs must engage with the brackets on the fuel tank and frame.

Sidepanels – TZR models
4 Each sidepanel is retained by a single screw and three pegs which push into grommets set in the frame and fuel tank. Ensure that the pegs locate fully with their grommets on refitting.

Sidepanels – DT models
5 The right-hand sidepanel is secured by a single screw, a peg at the front and a hook at the rear. The left-hand sidepanel is secured by a bolt at the rear and a peg at its front edge. Ensure that the pegs and hook are located correctly with the grommet on refitting.
6 The air scoops are secured by a single screw at the front and by two screws to the fuel tank.

Chapter 6 Wheels, brakes and tyres

Contents

General description	1
Front wheel: removal	2
Front wheel: refitting	3
Rear wheel: removal	4
Rear wheel: refitting	5
Wheel bearings: removal, examination and refitting	6
Rear sprocket: removal, examination and refitting	7
Cush drive: examination and renovation – TZR model only	8
Brake caliper: removal, overhaul and refitting	9
Front brake master cylinder: removal, overhaul and refitting	10
Rear brake master cylinder: removal, overhaul and refitting	11
Brake discs: examination and renewal	12
Bleeding the hydraulic brake system	13
Rear drum brake: examination and renovation – TZR 2RK and 3PC1 models	14
Tubed tyres: removal, repair and refitting	15
Tubeless tyres: removal and refitting	16
Tubeless tyres: puncture repair and tyre renewal	17
Tyre valves: general	18
Wheel balancing	19

Specifications

Wheels

	Front	Rear
Size:		
Early TZR models (2RK and 3PC1)	1.85 x 16	2.15 x 18
Later TZR models (3PC2 and 3PC3)	2.15 x 17	2.50 x 18
DT models	1.60 x 21	1.85 x 18
Rim runout – radial and axial	2 mm (0.08 in)	2 mm (0.08 in)

Disc brake

	Front	Rear
Disc thickness:		
Early TZR models (2RK and 3PC1)	4.0 mm (0.16 in)	Not applicable
Later TZR models (3PC2 and 3PC3)	5.0 mm (0.20 in)	5.0 mm (0.20 in)
DT models	3.5 mm (0.14 in)	4.5 mm (0.18 in)
Service limit:		
Early TZR models (2RK and 3PC1)	3.5 mm (0.14 in)	Not applicable
Later TZR models (3PC2 and 3PC3)	4.5 mm (0.18 in)	4.5 mm (0.18 in)
DT models	3.0 mm (0.12 in)	4.0 mm (0.16 in)
Disc maximum runout – front and rear:		
TZR models	0.50 mm (0.020 in)	
DT models	0.15 mm (0.006 in)	
Caliper bore ID:		
Early TZR models (2RK and 3PC1)	38.1 mm (1.50 in)	Not applicable
Later TZR models (3PC2 and 3PC3)	42.8 mm (1.69 in)	38.1 mm (1.50 in)
DT models	34.9 mm (1.37 in)	30.2 mm (1.19 in)
Master cylinder bore ID – front and rear	12.7 mm (0.50 in)	
Brake pad friction material thickness:		
Early TZR models (2RK and 3PC1)	5.5 mm (0.22 in)	Not applicable
Later TZR models (3PC2 and 3PC3)	5.3 mm (0.21 in)	5.5 mm (0.22 in)
DT models	6.0 mm (0.24 in)	6.0 mm (0.24 in)
Service limit – front and rear:		
TZR models	0.5 mm (0.02 in)	
DT models	0.8 mm (0.03 in)	
Brake fluid type	DOT 4 (if not available DOT 3 may be used)	

Drum brake

Drum ID	130 mm (5.12 in)	
Service limit	131 mm (5.16 in)	
Shoe lining thickness	4.0 mm (0.16 in)	
Service limit	2.0 mm (0.08 in)	
Return spring free length	36.5 mm (1.44 in)	

Chapter 6 Wheels, brakes and tyres

Tyres

	Front	Rear
Size:		
Early TZR models (2RK and 3PC1)	90/90-16 48P	100/90-18 56P
Later TZR models (3PC2 and 3PC3)	90/80-17 46S	100/90-18 56S
DT models	2.75 x 21 4PR	4.10 x 18 4PR

Tyre pressures .... See Routine maintenance specifications

Torque settings

Component	kgf m	lbf ft
Front wheel spindle:		
TZR model	7.4	53
DT model	5.8	42
Front wheel spindle pinch bolt – TZR model	1.5	11
Front wheel spindle clamp nuts – DT model	1.0	7.2
Rear wheel spindle nut:		
TZR model	8.5	61
DT model	9.0	65
Rear sprocket nuts/bolts:		
TZR model	5.0	36
DT model	3.5	25
Brake caliper mounting bolts – TZR model	3.5	25
Rear caliper bracket to swinging arm bolt – TZR model	3.0	22
Brake caliper body bolt – DT model	1.8	13
Swinging arm end screws – DT model	0.3	2.2
Front brake caliper bracket bolts – DT model	3.5	25
Front brake master cylinder clamp bolts	0.9	6.5
Rear brake master cylinder screws	1.0	7.2
Brake hose union bolts	2.6	19
Bleed nipples	0.6	4.3
Brake disc screws/bolts	2.0	14
Drum brake camshaft pinch bolt	0.8	5.8
Drum brake torque arm to backplate nut	1.9	13

1 General description

The TZR models are fitted with cast aluminium wheels. Earlier models (2RK and 3PC1) use a 16 inch diameter front wheel and an 18 inch diameter rear wheel; both carry tubed tyres. The front brake is an hydraulically operated disc and the rear is a drum brake, operated by a rod linkage from the brake pedal. Later TZR models (3PC2 and 3PC3) are fitted with a 17 inch front wheel and an 18 inch rear wheel, designed to accept tubeless tyres. Both front and rear brakes are hydraulically operated discs.

The DT models are fitted with wheels of conventional wire-spoked construction, using chromed steel rims of 21 inch diameter at the front, and 18 inch diameter at the rear. Both wheels are fitted with tubed tyres. Both front and rear brakes are hydraulically operated discs.

2 Front wheel: removal

1 Support the machine on a strong wooden box or similar support to hold the machine securely so that its front wheel is clear of the ground. On the TZR model this may require the removal of the lower fairing beforehand and possibly the exhaust system.

TZR model
2 Remove the two bolts which retain the brake caliper to the fork leg. Lift the caliper clear of the wheel and forks, tying it to the frame to avoid placing undue strain on the hydraulic hose. Before proceeding further, insert a wooden wedge between the brake pads; this will prevent the pistons being expelled if the brake lever is operated accidentally.
3 Disconnect the speedometer cable by releasing the knurled retaining ring and pulling the cable out of the speedometer gearbox. Loosen the pinch bolt situated on the right-hand fork leg. The wheel spindle can then be unscrewed and withdrawn and the wheel lowered to the ground and lifted clear of the forks.

DT model
4 Remove the two bolts which clamp the disc cover to the fork leg and remove the front half of the cover. Pull the speedometer cable from

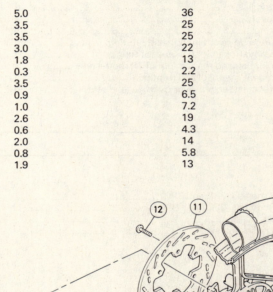

Fig. 6.1 Front wheel – TZR models (2RK and 3PC1 model shown – 3PC2 and 3PC3 similar)

1 Wheel spindle
2 Right-hand spacer
3 Grease seal
4 Bearing – 2 off
5 Spacer flange
6 Hub spacer
7 Speedometer driveplate
8 Retainer
9 Grease seal
10 Speedometer drive gearbox
11 Brake disc
12 Bolt – 6 off

Chapter 6 Wheels, brakes and tyres

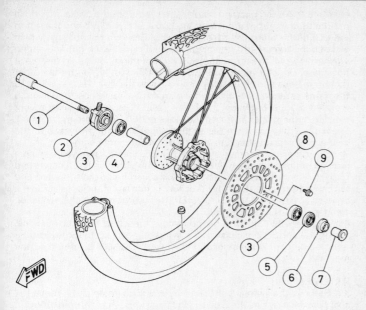

its gearbox after releasing the knurled retaining ring. The cable can be released from its holder on the right-hand fork leg if required, or can be taped to the fork leg clear of the wheel hub.

5 Slacken the four nuts on the wheel spindle retaining clamp to release the wheel spindle. Unscrew the spindle and withdraw it from the right-hand fork leg. The wheel can now be lowered to the ground and lifted clear of the forks. Before proceeding further, insert a wooden wedge between the brake pads. This will prevent the pistons being expelled if the brake lever is operated accidentally.

All models
6 Refer to Routine maintenance for details of wheel examination.

3 Front wheel: refitting

1 The wheel is refitted by reversing the removal sequence. Before fitting the wheel, grease the oil seal lips and speedometer drive gearbox with a lithium-based grease. Make sure that, when fitting the speedometer drive gearbox to the wheel, the drive plate tangs locate correctly in their slots in the hub. Refit the wheel spacers in their correct positions and, remove the wooden wedge from the brake caliper (DT model only at this stage).
2 Lift the wheel into position making sure the projection on the fork leg locates between the two lugs on the speedometer gearbox. On DT models also check the brake disc is correctly positioned in the caliper.
3 Insert the wheel spindle and tighten the nut by hand only at this stage. On TZR models remove the wooden wedge from between the brake pads and refit the brake caliper, tightening its mounting bolts to the specified torque setting. On DT models refit the front half of the disc cover.
4 Move the machine off its stand and with the front brake applied, compress the front forks several times to settle the components. Then tighten the wheel spindle and pinch bolt (or clamp nuts as appropriate) to the specified torque settings. On DT models tighten the top nuts of

Fig. 6.2 Front wheel – DT models

1 Wheel spindle
2 Speedometer drive gearbox
3 Bearing – 2 off
4 Hub spacer
5 Grease seal
6 Dust seal
7 Left-hand spacer
8 Brake disc
9 Bolt – 6 off

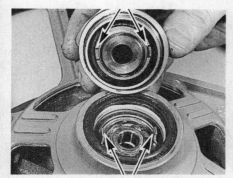

3.1a Refit speedometer drive gearbox, ensuring tangs locate in slots provided (arrowed), ...

3.1b ... and spacer – TZR model shown

3.2a Ensure protruding lug on fork leg is positioned between ribs on gearbox – TZR model

3.2b DT model – position disc correctly between brake pads whilst refitting wheel

3.4a TZR model – tighten the wheel spindle and pinch bolt to the specified torque setting

3.4b DT model – check spindle clamp is correctly fitted (arrow facing up) and tighten nuts as described in text

Chapter 6 Wheels, brakes and tyres

the spindle clamp first, then the lower two. Note that the gap between the fork leg and lower part of the clamp is intentional. Finally reconnect the speedometer cable and check the operation of the brakes and forks before taking the machine out on the road.

4 Rear wheel: removal

1 Raise the wheel clear of the ground by placing the machine on a strong wooden box, or similar support placed under the engine/gearbox unit. On the TZR model this may require the removal of the lower fairing and exhaust system.

TZR models
2 On early models (2RK and 3PC1), unscrew the brake adjuster nut from the brake rod end and displace the rod from the brake arm. Push out the trunnion from the end of the brake arm and fit it and the nut to the end of the brake rod for safekeeping. Remove the split pin which secures the torque arm retaining nut, then remove the nut and spring washer and pull the torque arm off its mounting bolt set in the brake backplate. On later models (3PC2 and 3PC3), remove both the caliper mounting bolts and lift the caliper away from the disc. Tie the caliper to the frame to avoid placing any strain on the hydraulic hose and insert a wooden wedge between the brake pads. The wedge will prevent ejection of the caliper pistons if the brake pedal is accidentally operated. Loosen the bolt which mounts the caliper mounting bracket to the swinging arm just enough to allow the bracket to move freely.
3 Straighten and remove the split pin which retains the wheel spindle nut and remove the nut. Release the locknuts on the chain adjusters and unscrew the drawbolts so that the wheel can be pushed forward and the chain disengaged from its sprocket. If there is insufficient slack in the drive chain to allow it to be lifted off the rear sprocket, the chain must be disconnected at its joining link.
4 Tap the wheel spindle out and remove the wheel from the machine, noting the correct positions of the wheel spacers so that they can be returned to their original locations when the wheel is refitted to the machine.

DT model
5 Remove the swinging arm end screws. Straighten and remove the split pin which retains the wheel spindle nut and slacken the spindle nut. Rotate both snail cam adjusters so that the wheel can be pushed forward and the chain disengaged from its sprocket. If there is insuffi-

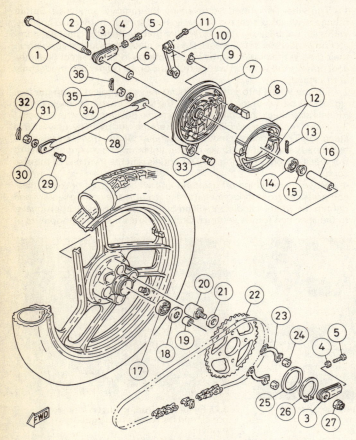

Fig. 6.3 Rear wheel – TZR (2RK and 3PC1) models

1 Wheel spindle
2 Split pin
3 Chain adjuster – 2 off
4 Locknut – 2 off
5 Adjuster bolt – 2 off
6 Right-hand spacer
7 Brake backplate
8 Brake camshaft
9 Wear indicator
10 Brake operating arm
11 Bolt
12 Brake shoes
13 Return spring – 2 off
14 Bearing
15 Spacer flange
16 Hub spacer
17 Bearing
18 Grease seal
19 Left-hand spacer
20 Cush drive bush – 4 off
21 Nylon spacer – 4 off
22 Sprocket
23 Tab washer – 2 off
24 Nut – 4 off
25 Washer
26 Circlip
27 Nut
28 Torque arm
29 Bolt
30 Spring washer
31 Nut
32 Split pin
33 Bolt
34 Spring washer
35 Nut
36 Split pin

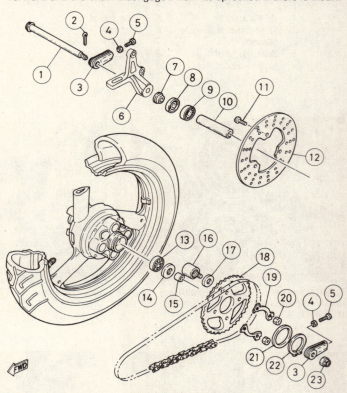

Fig. 6.4 Rear wheel – TZR (3PC2 and 3PC3) models

1 Wheel spindle
2 Split pin
3 Chain adjuster – 2 off
4 Locknut – 2 off
5 Adjuster bolt – 2 off
6 Caliper bracket
7 Right-hand spacer
8 Grease seal
9 Bearing
10 Hub spacer
11 Bolt – 3 off
12 Brake disc
13 Bearing
14 Grease seal
15 Left-hand spacer
16 Cush drive bush – 4 off
17 Nylon spacer – 4 off
18 Sprocket
19 Tab washer – 2 off
20 Nut – 4 off
21 Washer
22 Circlip
23 Nut

Chapter 6 Wheels, brakes and tyres

cient slack in the drive chain to allow it to be lifted off the rear sprocket, the chain must be disconnected at its joining link.
6 Remove the spindle nut, washer and chain adjuster and tap the wheel spindle out. The wheel can now be removed from the frame. Once the wheel is clear of the frame, place a wooden wedge between the brake pads to prevent ejection of the caliper pistons if the brake pedal is accidentally operated.

All models
7 Refer to Routine maintenance for details of wheel examination.

5 Rear wheel: refitting

TZR models
1 The wheel is installed by reversing the removal sequence. The chain adjusters must be fitted so that their stamped alignment marks can be seen aligning with those marks on the swinging arm. Refit the wheel spacers to the wheel, having first greased the oil seal lips. Check that the wheel spindle is straight, clean, and free from corrosion, then smear a liberal quantity of high melting-point grease over it to assist dismantling in the future. Install the chain over the rear sprocket and insert the wheel into the swinging arm fork ends. Push the spindle through, ensuring that the two chain adjusters and the two spacers are correctly fitted, then fit the spindle nut finger-tight. If disconnected remake the chain joining link, noting that the closed end of the spring clip must face the direction of normal wheel rotation. Check that the chain tension and wheel alignment are correct. Information on chain adjustment and wheel alignment can be found in Routine maintenance.
2 On the early models (2RK and 3PC1), once chain adjustment is correct, refit the brake torque arm, tightening by hand only its retaining nut. Reconnect the brake rod, apply firmly the rear brake to centralise the brake shoes and backplate on the drum, and tighten the rear wheel spindle nut to its recommended torque setting. Maintain pressure on the brake pedal and tighten the torque arm retaining nut to the specified

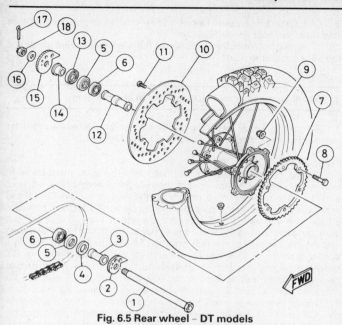

Fig. 6.5 Rear wheel – DT models

1 Wheel spindle
2 Chain adjuster
3 Left-hand spacer
4 Dust seal
5 Grease seal – 2 off
6 Bearing – 2 off
7 Sprocket
8 Bolt – 6 off
9 Nut – 6 off
10 Brake disc
11 Screw – 6 off
12 Hub spacer
13 Dust seal
14 Right-hand spacer
15 Chain adjuster
16 Washer
17 Split pin
18 Nut

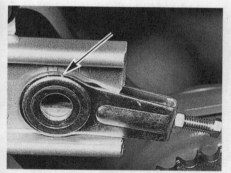

5.1a Ensure chain adjusters are fitted with alignment marks facing outwards (arrowed)

5.1b Do not omit wheel spacers

5.1c On drum brake models install right-hand spacer whilst inserting wheel spindle

5.2a Tighten spindle nut to specified torque setting and secure with a new split pin

5.2b On drum brake models secure torque arms bolts with R-pins

5.6 Ensure snail cam adjusters are fitted as shown

Chapter 6 Wheels, brakes and tyres

torque setting. Secure both nuts by fitting an R-pin through each. Adjust the rear brake and stop lamp switch as described in Routine maintenance.

3 On later models (3PC2 and 3PC3) tighten the wheel spindle nut to its recommended torque setting and fit a new split pin. Remove the wooden wedge from between the brake pads and refit the caliper to its mounting bracket, tightening all the mounting bolts to the torque settings given in the Specifications at the start of this Chapter.

4 On all models, thoroughly check the operation of the rear brake before taking the machine out on the road.

DT model

5 The wheel is installed by reversing the removal sequence. Smear a small quantity of lithium-based grease over the dust seal lips and insert the wheel spacers, noting that they should be fitted in the correct positions: see accompanying illustration.

6 Check that the wheel spindle is straight, clean, and free from corrosion, then smear a liberal quantity of high melting-point grease over it to assist dismantling in the future and slide the left-hand snail cam along it to fit against the spindle head, making sure that the wheel alignment marks are facing outwards. Insert the spindle into the left-hand fork of the swinging arm, and remove the wooden wedge from between the brake pads. Unless previously disconnected, install the chain over the rear sprocket whilst installing the wheel, complete with spacers, into the swinging arm and manoeuvring it into position. Make sure that the brake disc is correctly positioned in the caliper and then push the spindle fully into position. Fit the right-hand snail cam adjuster and plain washer. Secure with the nut, tightening it by hand only at this stage. Refit the screws in the ends of the swinging arm, securing their nuts finger-tight.

7 If previously disconnected remake the final drive chain, noting that the closed end of the spring clip joining link should face the direction of normal wheel rotation. Adjust the chain and wheel alignment as described in Routine maintenance.

8 Tighten the wheel spindle nut and swinging arm end screws to the specified torque settings, and fit a new split pin to the spindle nut.

9 Check the brake pedal height and stop lamp switch operation as described in Routine maintenance.

10 Thoroughly check the operation of the rear brake before taking the machine out on the road.

6 Wheel bearings: removal, examination and refitting

1 Before the wheel bearings can be examined, the wheel must be removed as described in Section 2 or 4 of this Chapter.

Front wheel

2 On the TZR model, the speedometer drive gearbox must be withdrawn, followed by the grease seal, the speedometer drive ring retainer, and the speedometer drive ring. Heat the end of an old flat-bladed screwdriver and bend the tip into a slightly curved shape with no sharp edges. This will give a useful tool for levering out the oil seal without damaging the sealing lip or the spring beneath it. The spacer and dust seal on the disc side of the hub can then be withdrawn, if desired, in a similar manner, or they can be driven out with the bearings. For DT models the speedometer drive gearbox must be withdrawn. The grease seal, dust cover and spacer on the opposite side of the hub can be removed as described above or driven out with the bearings as desired.

3 Support the wheel firmly on two wooden blocks placed as close to the hub centre as possible to prevent distortion, ensuring that enough space is allowed to permit bearing removal. Place the end of a small flat-ended drift against the upper face of the lower bearing and tap the bearing downwards out of the wheel hub. The spacer located between the two bearings may be moved sideways slightly in order to allow the drift to be positioned against the face of the bearing. Move the drift around the face of the bearing whilst drifting it out of position, so that the bearing leaves the hub squarely.

4 With the one bearing removed, the wheel may be lifted and the spacer withdrawn from the hub. Invert the wheel and remove the second bearing, using a similar procedure to that used for the first.

6.8a Fit first wheel bearing then invert wheel and fit central spacer – grease cavity ...

6.8b ... and fit second bearing ...

6.8c ... using a hammer and tubular drift

6.9a TZR model – fit speedometer drive ring as shown ...

6.9b ... followed by drive ring retainer

6.9c Grease seal to aid installation

Chapter 6 Wheels, brakes and tyres

5 Wash the bearings thoroughly in a high flash-point solvent to remove all traces of the old grease. Check the bearing tracks and balls for wear or pitting or damage to the hardened surfaces. A small amount of side movement in the bearing is normal but no radial movement should be detectable. Check the bearings for play and roughness when they are spun by hand. All used bearings will emit a small amount of noise when spun but they should not chatter or sound rough. If there is any doubt about the condition of the bearings they should be renewed.
6 Carefully clean the bearing recesses in the hub and the centre cavity. All traces of the old grease, which may be contaminated with dirt, must be removed. Examine the grease seals and renew them if any damage or wear is found.
7 Before replacing the bearings pack them with lithium-based, high melting-point grease. This applies equally to the original bearings, if refitted, and to new ones, if the originals are to be renewed. With the wheel firmly supported on the two wooden blocks, tap a bearing into place in the hub. Use a hammer and a tubular metal drift or socket spanner *which bears only on the outer race* of the bearing to drive the bearing into position.
8 Once one bearing has been installed, invert the wheel, fit the central spacer and pack the remaining space no more than ⅔ full of lithium-based, high melting-point grease. Once the grease is packed in, invert the wheel and fit the second bearing in the same manner as the first.
9 On TZR models the speedometer drive ring should now be fitted, to the left-hand side of the hub, ensuring that the drive tangs are located in the slots provided for them. Insert the drive ring retainer. On all models refit the grease seals to each side of the hub, noting that if the seals were damaged on removal, or if there is any doubt about their condition, they should be renewed. Use a hammer and a tubular drift or socket spanner which bears on the outside diameter of the seal to drive it gently into position. On DT125 models press the dust cover into position in the hub left-hand side. Install the speedometer drive gearbox, aligning the drive plate tangs correctly.
10 Fit the front wheel assembly into the front forks as described in Section 3.

Rear wheel
11 Remove the rear wheel as described in Section 4, then withdraw the brake backplate (where fitted) and the wheel spacer(s) to gain access to the rear wheel bearings. Due to the similarity in design and construction between the front and rear hubs of each of the machines described in this Manual, the procedures for removal, examination, and refitting of the rear wheel bearings are exactly the same as those given above.

7 Rear sprocket: removal, examination and refitting

1 Examine the teeth of the rear sprocket. If these are hooked, chipped, or otherwise damaged, the sprocket must be renewed. Note that it is considered bad practice to renew just one sprocket or the chain alone; both front and rear sprockets and the chain should be renewed together at all times.

TZR model
2 Remove the rear wheel from the machine as described in Section 4. Withdraw the brake backplate assembly (where fitted) and lay the wheel on a convenient working surface with the sprocket uppermost. It is useful to place a sheet of cardboard or several layers of newspaper on the working surface to protect the wheel's finish. Using a suitable pair of circlip pliers, remove the circlip which retains the sprocket on the hub and lift away the washer immediately below it. Bend back the locking tabs of the tab washers, then slacken and remove the four nuts which secure the cush drive rubbers to the sprocket. The sprocket can then be lifted off the hub.
3 When refitting the sprocket always renew the tab washers and the circlip. Tighten the four retaining nuts in a diagonal sequence to the specified torque setting.

DT model
4 Remove the rear wheel from the machine as described in Section 4. The sprocket is mounted directly on the wheel hub left-hand face by six bolts and self-locking nuts. Before fitting the sprocket, make sure the surfaces of the hub which come in contact with the sprocket are clean.

7.2a TZR model – remove sprocket retaining nuts and tab washers ...

7.2b ... followed by large circlip ...

7.2c ... and washer ...

7.2d ... and lift off sprocket

7.3 Tighten nuts to specified torque setting and secure with tab washer as shown

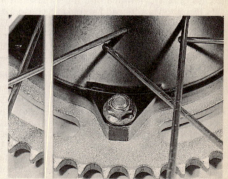
7.4 On DT model, sprocket is retained by six bolts and self-locking nuts

Chapter 6 Wheels, brakes and tyres

8.4a Cush drive bushes can be driven out via access holes in opposite side of hub ...

8.4b ... using a hammer and drift – 2RK and 3PC1 model arrangement shown

8.5 Do not omit nylon spacers

Insert the sprocket mounting bolts from the outside with the self-locking nuts positioned on the inside of the hub. Note that the manufacturer recommends that a few drops of thread-locking compound, such as Loctite, be applied to the bolt threads. Prior to reassembly inspect the retaining nuts; if they have lost their locking action they must be renewed. Tighten the bolts in a diagonal sequence to the specified torque setting.

5 Refit the wheel to the machine as described in Section 5.

8 Cush drive: examination and renovation – TZR model only

1 The cush drive assembly consists of four tubular bushes located in the hub. These bushes give a cushioning effect to the drive sprocket and drive.
2 To gain access to the bushes, the sprocket has to be removed as described in Section 7. Renewal of the bushes is required when there is excessive free play at the sprocket, often accompanied by snatch in the final drive.
3 If the cush drive rubbers are to be renewed, the first step is to apply a liberal dose of penetrating fluid to each cush drive bush, removing the nylon ring from each to permit this, and to leave the wheel for as long as possible to allow the fluid to work. When ready, invert the wheel and place it on the work surface on top of two wooden blocks placed as close around the hub as possible to give maximum support. The blocks must be thick enough for the wheel to be held at a height that will permit the removal of the bushes.
4 Access to the rear of the bushes can be gained through holes in the opposite side of the wheel hub. Find the largest drift available which will fit through the access holes, and using a suitably heavy hammer, drive out the bush with a few healthy blows. When the bushes have been removed thoroughly clean the bush recesses in the hub, removing all burrs, scratches and any traces of corrosion.
5 To fit the new bushes apply a thin smear of grease to both the bush and recess. This will aid refitting and prevent corrosion. Tap each bush firmly into place using a suitably sized socket which bears only on the metal outer of the bush. When all four bushes are fitted, make sure all excess grease is removed from the brake drum (earlier models). Then refit the nylon spacers and sprocket as described in Section 7.

9 Brake caliper: removal, overhaul and refitting

1 Remove the two bolts which retain the front disc and caliper covers and remove both covers (DT models only). Disconnect the union bolt from the brake caliper and drain the hydraulic fluid into a suitable container. At this stage, it is as well to stop the flow of fluid from the reservoir, by holding the front lever in against the handlebar. This is easily done using a stout elastic band, or alternatively, a section cut from an old inner tube.
2 **Note:** Brake fluid will discolour or remove paint if contact is allowed. Avoid this where possible and remove accidental splashes immediately. Similarly, avoid contact between the fluid and plastic parts such as instrument lenses, as damage will also be done to these. When all the fluid is drained from the hose, clean the connections carefully and secure the hose end and fittings inside a clean polythene bag, to await reassembly. As with all hydraulic systems, it is most important to keep each component scrupulously clean, and to prevent the ingress of any foreign matter. For this reason, it is as well to prepare a clean area in which to work, before further dismantling. As in any form of component dismantling, ensure that the outside of the caliper is thoroughly cleaned down.
3 On TZR models, remove the brake caliper from the machine and remove the brake pads from the caliper as described in Routine maintenance.
4 On DT models, the brake pads must be removed before the caliper can be removed from the machine, again refer to Routine maintenance. Once the pads have been removed, the caliper body can be parted from its mounting bracket. Disconnect the rubber boot from the guide pin on the bracket and pull the caliper away. The caliper is now ready for overhaul.

DT model

5 The piston may be driven out of the caliper body by an air jet – a foot pump if necessary. Under no circumstances should any attempt be made to lever or prise the piston out of the caliper. If the compressed air method fails, temporarily reconnect the caliper to the flexible hose, and use the handlebar lever to displace the piston hydraulically. Wrap some rag around the caliper to catch the inevitable shower of brake fluid. Carefully remove the circlip (front caliper only) which retains the dust and fluid seals and remove both seals.
6 Clean each part carefully, using only clean hydraulic fluid. On no account use petrol, oil or paraffin as these will cause the seals to degrade and swell. Keep all components dust free. Examine the piston surface for scoring or pitting – any imperfection will necessitate renewal. The seals should be renewed as a matter of course, re-using an old seal is a false economy. Remember that the safety of the machine is very much dependent on seal and piston condition.
7 Reassemble the caliper, again ensuring absolute cleanliness, by reversing the dismantling procedure. Use clean hydraulic fluid to lubricate the seals before fitting the piston.
8 Before refitting the caliper, check the caliper mounting bracket for cracks or damage, also check the guide pin and rubber boots for corrosion and deterioration, if any component is damaged in any way it should be renewed. The front mounting bracket can be released by removing the two bolts which retain it to the fork leg. To remove the rear caliper mounting bracket, it is first necessary to remove the rear wheel as described in Section 4. Refit the bracket by reversing the removal sequence and tighten the mounting bolts (front) to their specified torque setting. Lubricate the guide pin on the caliper mounting bracket with a smear of lithium-based grease before installing the caliper.
9 Refit the pads and springs as described in Routine maintenance. Make sure that the rubber boots on the guide pin and caliper bolt are correctly located once the caliper is in position, as they prevent the ingress of moisture and the subsequent build-up of corrosion which could lead to the caliper seizing.

TZR model

10 To displace the pistons it is necessary to hold one in place while the other is expelled using compressed air. Yamaha recommend that the right-hand piston is held in place using a pair of slip-joint pliers with rag to protect the piston surface. Compressed air is then applied to the fluid inlet to expel the piston. When work on the first piston is complete,

Chapter 6 Wheels, brakes and tyres

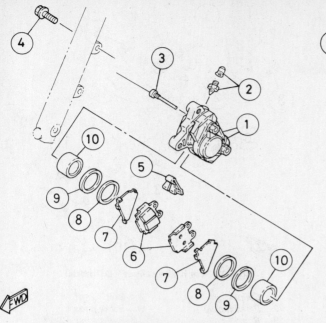

Fig. 6.6 Front brake caliper – TZR (2RK and 3PC1) models

1. Caliper
2. Bleed nipple
3. Pad retaining pin – 2 off
4. Bolt – 2 off
5. Spring
6. Brake pads
7. Shim – 2 off
8. Dust seal – 2 off
9. Fluid seal – 2 off
10. Piston – 2 off

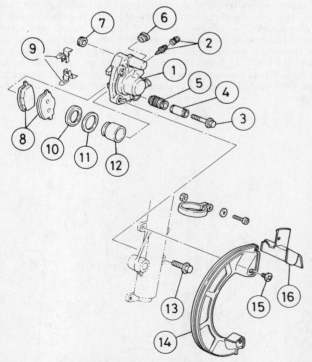

Fig. 6.8 Front brake caliper – DT models

1. Caliper
2. Bleed nipple
3. Bolt
4. Sleeve
5. Rubber boot
6. Pad inspection plug
7. Rubber boot
8. Brake pads
9. Pad spring – 2 off
10. Dust seal
11. Fluid seal
12. Piston
13. Bolt – 2 off
14. Disc cover
15. Screw – 2 off
16. Caliper cover

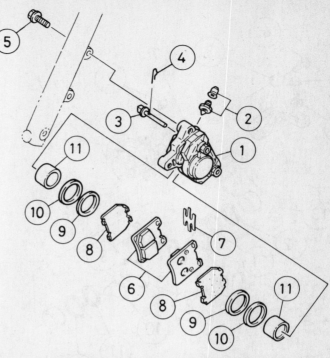

Fig. 6.7 Front brake caliper – TZR (3PC2 and 3PC3) models

1. Caliper
2. Bleed nipple
3. Pad retaining pin – 2 off
4. R-clip – 2 off
5. Bolt – 2 off
6. Brake pads
7. Pad spring
8. Shim – 2 off
9. Dust seal – 2 off
10. Fluid seal – 2 off
11. Piston – 2 off

refit it and then expel the remaining piston in a similar manner. In the absence of compressed air, temporarily refit the hydraulic hose, and gradually "pump" the pistons out of their bores. Whichever method is employed take great care to avoid getting fingers trapped by the emerging pistons. Note: **on no account** slacken or remove the caliper bridge bolts which secure the two halves of the caliper body; these do not need to be separated during overhaul.

11 Examine and clean the caliper components as described in paragraph 6 above, remembering not to interchange the pistons. If both are removed from the caliper at the same time it is advisable to clean them with methylated spirit and then to mark them L and R using a spirit-based marker pen on the recessed outer face of each one. Reassembly is a reversal of the removal sequence noting that absolute cleanliness is vital. Always fit new seals regardless of apparent condition, and lubricate them with hydraulic fluid during installation.

12 Refit the brake pads and shims to the caliper as described in Routine maintenance and install the caliper on the machine.

All models

13 Refit the union bolt using new copper sealing washers on each side of the hose union and tighten to the specified torque setting. Remember the system will now require filling with the recommended brake fluid and bleeding before use; refer to Section 13 of this Chapter. Finally refit the disc cover (DT models), and check the brake operation thoroughly before taking the machine on the road.

10 Front brake master cylinder: removal, overhaul and refitting

1 The master cylinder forms a unit with the hydraulic fluid reservoir and front brake lever, and is mounted by a clamp to the right-hand side of the handlebars.

Chapter 6 Wheels, brakes and tyres

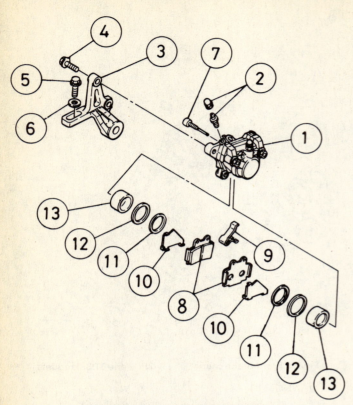

Fig. 6.9 Rear brake caliper – TZR models

1. Caliper
2. Bleed nipple – 2 off
3. Mounting bracket
4. Bolt – 2 off
5. Bolt
6. Washer
7. Pad retaining pin – 2 off
8. Brake pads
9. Spring
10. Shim – 2 off
11. Dust seal – 2 off
12. Fluid seal – 2 off
13. Piston – 2 off

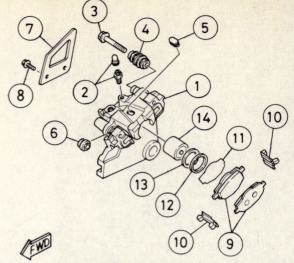

Fig. 6.10 Rear brake caliper – DT model

1. Caliper
2. Bleed nipple
3. Bolt
4. Rubber boot
5. Pad inspection plug
6. Rubber boot
7. Guard plate
8. Bolt – 2 off
9. Brake pads
10. Pad spring – 2 off
11. Shim
12. Dust seal
13. Fluid seal
14. Piston

8 Reassemble carefully, using hydraulic fluid as a lubricant on seals and piston, reversing the dismantling sequence. Make sure the circlip and dust seal are fitted correctly, and that the unit is clamped securely to the handlebars. The clamp is marked 'UP' to assist correct refitting and its bolts must be tightened to the specified torque setting, upper bolt first. Reconnect the hydraulic hose, using new sealing washers each side of the union. Tighten the union bolt to the recommended torque setting. Refill the reservoir remembering to top up after the system has been bled by following the procedure given in Section 13 of this Chapter.

2 The unit must be drained before any dismantling can be undertaken. Place a suitable container below the caliper unit and run a length of plastic tubing from the caliper bleed screw to the container. Unscrew the bleed screw one full turn and proceed to empty the system by squeezing the front brake lever. When all the fluid has been expelled, tighten the bleed screw and remove the tube.

3 Select a suitable clean area in which the various components may be safely laid out, a large piece of white lint-free cloth or white paper being ideal.

4 Remove the locknut and brake lever pivot to free the lever. As it is lifted away note the small spring which is fitted into the end of the lever blade. On DT models the hand protector guard can also be removed. Remove the brake switch by pressing the switch hook with a suitable tool and pulling it out of the cylinder.

5 Remove the two bolts which hold the master cylinder clamp half to the body and then lift the master cylinder away. Remove the cover and diaphragm, then empty the reservoir. If it is still connected, remove the union bolt and free the hydraulic hose from the master cylinder body. Tape a polythene bag over the end of the hydraulic hose to prevent the entry of dirt and leakage of brake fluid.

6 Pull off the dust seal from the end of the piston bore to expose the piston end and the circlip which retains it. Remove the circlip to free the piston. If the piston tends to stick in the bore it can be pulled clear using pointed-nose pliers. As the piston is removed the main seal and spring will be released.

7 Examine the piston and seals for scoring or wear and renew if imperfect. Excessive scoring may be due to contaminated fluid, and if this is suspected, it is probably worth checking the condition of the caliper seals and piston. Note that the master cylinder seals, piston and spring are supplied as a set – they cannot be obtained individually.

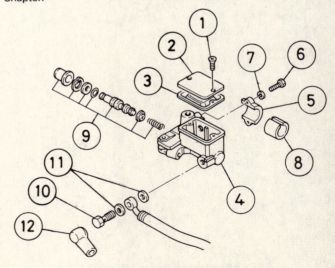

Fig. 6.11 Front brake master cylinder

1. Screw – 2 off
2. Cover
3. Diaphragm
4. Reservoir
5. Clamp
6. Bolt – 2 off
7. Spring washer – 2 off
8. Insert (DT only)
9. Piston assembly (DT shown – TZR similar)
10. Union bolt
11. Sealing washer – 2 off
12. Cover

Chapter 6 Wheels, brakes and tyres

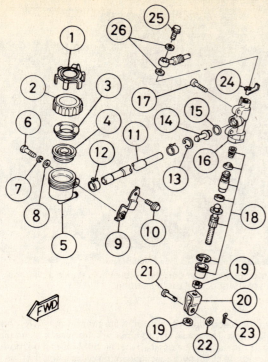

Fig. 6.12 Rear brake master cylinder – TZR model

1	Cover retainer	14	Hose union
2	Cover	15	O-ring
3	Diaphragm plate	16	Master cylinder
4	Diaphragm	17	Screw – 2 off
5	Reservoir	18	Piston assembly
6	Screw	19	Nut – 2 off
7	Spring washer	20	Brake pedal link
8	Washer	21	Cotter pin
9	Union	22	Washer
10	Bolt	23	Split pin
11	Reservoir hose	24	Special washer
12	Hose clamp – 2 off	25	Union bolt
13	Circlip	26	Sealing washer – 2 off

9 Thoroughly test the braking system before taking the machine on the road.

11 Rear brake master cylinder: removal, overhaul and refitting

1 The rear brake master cylinder can be dealt with in much the same way as has been described in the previous Section, the main differences arising in the procedure for removing the assembly from the frame. Drain the hydraulic fluid as described previously.

TZR model

2 Disconnect both the stoplamp and rear brake pedal return springs from the brake pedal. Remove the two screws which mount the master cylinder to the footrest bracket and the rear brake pedal retaining bolt. Pull the brake pedal off its pivot on the footrest bracket and remove the bracket from the frame. Disconnect the rear brake pedal by removing the split pin and withdrawing the cotter pin and washer from the forked end of the master cylinder.

3 Remove the union bolt from the top of the master cylinder and disconnect the hydraulic brake hose. Tape a polythene bag over the end of the hose to prevent the entry of dirt and the leakage of brake fluid. Note the position of the special washer before removing it from the top of the master cylinder so that it can be replaced correctly on reassembly. Before disconnecting the reservoir hose from the master cylinder find a small container to drain the brake fluid into. Alternatively the hose can be plugged with a suitably sized clean screw or bolt. Remove the hose clamp and disconnect the reservoir hose using one of the above methods to prevent the spillage of brake fluid. The master cylinder can now be dismantled and overhauled using the procedure in Section 10, paragraphs 6 to 8.

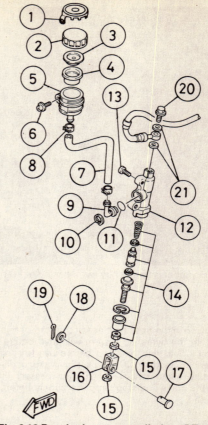

Fig. 6.13 Rear brake master cylinder – DT models

1	Cover retainer	12	Master cylinder
2	Cover	13	Screw – 2 off
3	Diaphragm plate	14	Piston assembly
4	Diaphragm	15	Nut – 2 off
5	Reservoir	16	Brake pedal link
6	Screw	17	Cotter pin
7	Reservoir hose	18	Washer
8	Hose clamp – 2 off	19	Split pin
9	Union	20	Union bolt
10	Circlip	21	Sealing washer – 2 off
11	O-ring		

DT model

4 Remove the rear brake hose guides situated on the swinging arm. Remove the union bolt from the top of the master cylinder and disconnect the brake hose. Tape a polythene bag over the end of the hose to prevent the entry of dirt and leakage of brake fluid. Disconnect the rear brake pedal by removing the split pin and withdrawing the cotter pin and washer from the forked end of the master cylinder. Remove the screw which retains the reservoir to the frame tube and also the two master cylinder mounting bolts. The master cylinder and reservoir assembly can now be manoeuvred out of the frame for inspection.

5 Unscrew the reservoir cap, remove the inner cap and diaphragm and empty the brake fluid into a suitable container. Then disconnect the brake hose from the master cylinder, the master cylinder can now be dismantled and overhauled using the procedure in Section 10, paragraphs 6 to 8.

All models

6 The master cylinder is refitted to the frame by a reverse of the removal sequence. Use a new split pin on the master cylinder cotter pin and renew both copper sealing washers on the union bolt. Make sure the union and all mounting bolts are tightened to their recommended

11.4 Rear master cylinder is retainer by two Allen bolts – DT model shown

12.3 Tighten brake disc mounting bolts evenly and in a diagonal sequence to the specified torque setting

torque settings given in the Specifications at the start of this Chapter.

7 Remember that the hydraulic system will now require filling with the specified brake fluid and bleeding before use, by following the instructions given in Section 13. Check the brake pedal height as described in Routine maintenance and adjust if necessary. Finally check the operation of the brake and stop lamp thoroughly before taking the machine on the road.

12 Brake discs: examination and renewal

1 Examination of the discs can be carried out with the wheels installed. Look for signs of excessive scoring. Some degree of scoring is inevitable, but in severe cases renewal of the disc may prove necessary to restore full braking effect. Check the disc for warpage, which can often result from overheating or impact damage and may cause brake judder. This is best checked using a dial gauge mounted on the fork leg, or swinging arm and should not exceed the specified limit.
2 The disc thickness should be measured using a vernier caliper or micrometer in several places around the disc surface. If the disc has worn to or beyond the service limit it should be renewed.
3 To remove the disc, first remove the relevant wheel as described in this Chapter, then unscrew the screws or bolts evenly and in a diagonal sequence. When refitting the disc check that the mounting surfaces are clean and apply Loctite to the mounting bolts or screws. Tighten them evenly and progressively to the specified torque setting.

13 Bleeding the hydraulic brake system

1 The method of bleeding a brake system of air and the procedure described below apply equally to either a front brake or rear brake of the hydraulically actuated type.
2 If the brake action becomes spongy, or if any part of the hydraulic system is dismantled (such as when a hose is renewed) it is necessary to bleed the system in order to remove all traces of air. The procedure for bleeding the hydraulic system is best carried out by two people.
3 Check the fluid level in the reservoir and top up with new fluid of the specified type if required. Keep the reservoir at least half full during the bleeding procedure; if the level is allowed to fall too far air will enter the system requiring that the procedure be started again from scratch. Screw the cap onto the reservoir to prevent the ingress of dust or the ejection of a spout of fluid.
4 Remove the dust cap from the caliper bleed nipple and clean the area with a rag. Place a clean glass jar below the caliper and connect a pipe from the bleed nipple to the jar. Note that on the TZR model rear brake caliper it will be necessary to run a pipe from each bleed nipple. A clear plastic tube should be used so that air bubbles can be more easily seen. Place some clean hydraulic fluid in the glass jar so that the pipe is immersed below the fluid surface throughout the operation.
5 If parts of the system have been renewed, and thus the system must be filled, open the bleed nipple about one turn and pump the brake lever until fluid starts to issue from the clear tube. Tighten the bleed nipple and then continue the normal bleeding operation as described in the following paragraphs. Keep a close check on the reservoir level whilst the system is being filled.
6 Operate the brake lever/pedal as far as it will go and hold it in this position against the fluid pressure. If spongy brake operation has occurred it may be necessary to pump rapidly the brake lever/pedal a number of times until pressure is achieved. With pressure applied, loosen the bleed nipple about half a turn. Tighten the nipple as soon as the lever/pedal has reached its full travel and then release the lever/pedal. Repeat this operation until no more air bubbles are expelled with the fluid into the glass jar. When this condition is reached the air bleeding operation should be complete, resulting in a firm feel to the brake operation. If sponginess is still evident continue the bleeding operation; it may be that an air bubble trapped at the top of the system has yet to work down through the caliper.

13.4 Connect bleed tube device to nipple as shown

Chapter 6 Wheels, brakes and tyres

14.2a Brake shoe lining thickness measurement

14.2b Drum surface must be free from brake dust and scoring

14.3 Fold brake shoes together in a 'V' to permit removal

7   When all traces of air have been removed from the system, top up the reservoir and refit the diaphragm and cap or cover, as appropriate. Check the entire system for leaks, and check also that the brake system in general is functioning efficiently before using the machine on the road.

8   Brake fluid drained from the system will almost certainly be contaminated, either by foreign matter or more commonly be the absorption of water from the air. All hydraulic fluids are to some degree hygroscopic, that is, they are capable of drawing water from the atmosphere, and thereby degrading their specifications. In view of this, and the relative cheapness of the fluid, old fluid should always be discarded.

9   Great care should be taken not to spill hydraulic fluid on any painted cycle parts; it is a very effective paint stripper. Also, the plastic glasses in the instrument heads, and most other plastic parts, will be damaged by contact with this fluid.

14 Rear drum brake: examination and renovation – TZR 2RK and 3PC1 models

1   When the wheel has been removed, as described in Section 4, the brake backplate assembly can be lifted away. Remove any accumulated dust from the brake drum by wiping it with a rag soaked in solvent. **Do not** blow the drum out with compressed air. The dust contains asbestos particles which are harmful if inhaled.

2   Measure the lining thickness at its thinnest point and renew the shoes if they are worn to or beyond the service limit shown in the Specifications Section. The shoes should also be renewed if they have become contaminated with oil or grease. The lining material is bonded to the shoes and this means that the shoes should be renewed complete.

3   Removal of the brake shoes is accomplished by folding the shoes together so that they form a 'V'. With the spring tension relaxed, both shoes and springs may be removed from the brake backplate as an assembly. Detach the springs from the shoes and carefully inspect them for any signs of fatigue or failure. Compare their free length with the figure given in the Specifications.

4   Before fitting the brake shoes, check that the brake operating cam is working smoothly and is not binding in its pivot. The cam can be removed by withdrawing the pinch bolt on the operating arm and pulling the arm off the shaft. Before removing the arm, it is advisable to mark its position in relation to the shaft, so that it can be relocated correctly, with the wear indicator pointer.

5   Remove any deposits of hardened grease or corrosion from the bearing surface of the brake cam and shoe by rubbing it lightly with a strip of fine emery paper or by applying solvent with a piece of rag. Lightly grease the length of the shaft and the face of the operating cam prior to reassembly. Clean and apply a smear of high melting-point grease to the pivot stub which is set in the backplate. Relocate the camshaft, having applied a smear of high melting-point grease to its shank. Align and fit the operating arm with the wear indicator pointer. Tighten the pinch bolt to the specified torque setting.

6   Fitting the brake shoes and springs to the brake backplate is a reversal of the removal procedure. Some patience will be needed to align the assembly with the pivot and operating cam whilst still retaining the spring in position; once they are correctly aligned, they can be pushed back into position by pressing downwards to snap them into position.

7   Examine the drum surface for signs of scoring, wear beyond the service limit or oil contamination. All of these conditions will impair braking efficiency. Remove all traces of dust, preferably using a brass wire brush, taking care not to inhale any of it, as it is of an asbestos

14.5a Apply a smear of high melting-point grease to brake camshaft prior to refitting ...

14.5b ... and install wear indicator pointer and operating arm

nature and consequently harmful. Remove oil or grease deposits, using solvent. If deep scoring is evident, due to the linings having worn through to the shoe at some time, the drum must be skimmed on a lathe, or renewed. Whilst there are firms who will undertake to skim a drum whilst it is fitted to the wheel, it should be borne in mind that excessive skimming will change the radius of the drum in relation to the brake shoes, therefore reducing the friction area until extensive bedding in has taken place. Also full adjustment of the shoes may not be possible. If in doubt about this point, the advice of one of the specialist engineering firms who undertake this work should be sought.

8 Refit the brake backplate assembly to the wheel and refit the wheel in the frame, using the information in Section 5.

15 Tubed tyres: removal, repair and refitting

1 To remove the tyre from either wheel, first detach the wheel from the machine. Deflate the tyre by removing the valve core, and when the tyre is fully deflated, push the bead away from the wheel rim on both sides so that the bead enters the centre well of the rim. Remove the locking ring and push the tyre valve into the tyre itself.

2 Insert a tyre lever close to the valve and lever the edge of the tyre over the outside of the rim. Very little force should be necessary; if resistance is encountered it is probably due to the fact that the tyre beads have not entered the well of the rim, all the way round. On TZR models note that damage to the soft alloy by tyre levers can be prevented by the use of plastic rim protectors.

3 Once the tyre has been edged over the wheel rim, it is easy to work round the wheel rim, so that the tyre is completely free from one side. At this stage the inner tube can be removed.

4 Now working from the other side of the wheel, ease the other edge of the tyre over the outside of the wheel rim that is furthest away. Continue to work around the rim until the tyre is completely free from the rim.

5 If a puncture has necessitated the removal of the tyre, reinflate the inner tube and immerse it in a bowl of water to trace the source of the leak. Mark the position of the leak, and deflate the tube. Dry the tube, and clean the area around the puncture with solvent. When the surface has dried, apply rubber solution and allow this to dry before removing the backing from the patch, and applying the patch to the surface.

6 It is best to use a patch of a self-vulcanizing type, which will form a permanent repair. Note that it may be necessary to remove a protective covering from the top surface of the patch after it has sealed into position. Inner tubes made from a special synthetic rubber may require a special type of patch and adhesive, if a satisfactory bond is to be achieved.

7 Before refitting the tyre, check the inside to make sure that the article that caused the puncture is not still trapped inside the tyre. Check the outside of the tyre, particularly the tread area to make sure nothing is trapped that may cause a further puncture.

8 If the inner tube has been patched on a number or past occasions, or if there is a tear or large hole, it is preferable to discard it and fit a replacement. Sudden deflation may cause an accident, particularly if it occurs with the rear wheel.

9 To refit the tyre, inflate the inner tube for it just to assume a circular shape but only to that amount, and then push the tube into the tyre so that it is enclosed completely. Lay the tyre on the wheel at an angle, and insert the valve through the rim tape (DT models only) and the hole in the wheel rim. Attach the locking ring on the first few threads, sufficient to hold the valve captive in its correct location.

10 Starting at the point furthest from the valve, push the tyre bead over the edge of the wheel rim until it is located in the central well. Continue to work around the tyre in this fashion until the whole of one side of the tyre is on the rim. It may be necessary to use a tyre lever during the final stages.

11 Make sure there is no pull on the tyre valve and, again commencing with the area furthest from the valve, ease the other bead of the tyre over the edge of the rim. Finish with the area close to the valve, pushing the valve up into the tyre until the locking ring touches the rim. This will ensure that the inner tube is not trapped when the last section of bead is edged over the rim with a tyre lever.

12 Check that the inner tube is not trapped at any point. Reinflate the inner tube, and check that the tyre is seating correctly around the wheel rim. There should be a thin rib moulded around the wall of the tyre on both sides, which should be an equal distance from the wheel rim at all points. If the tyre is unevenly located on the rim, try bouncing the wheel when the tyre is at the recommended pressure. It is probable that one of the beads has not pulled clear of the centre well.

13 Always run the tyres at the recommended pressures and never under- or over-inflate. The correct pressures are given in Routine maintenance.

14 Tyre refitting is aided by dusting the side walls, particularly in the vicinity of the beads, with a liberal coating of french chalk. Washing-up liquid can also be used to good effect, but this has the disadvantage, where steel rims are used, of causing the inner surface of the wheel rim to rust.

15 Never refit the inner tube and tyre without the rim tape in position (DT models only). If this precaution is overlooked there is a good chance of the ends of the spoke nipples chafing the inner tube and causing a crop of punctures.

16 Never fit a tyre that has a damaged tread or sidewalls. Apart from legal aspects, there is a very great risk of a blowout, which can have very serious consequences on a two wheeled vehicle.

17 Tyre valves rarely give trouble, but it is always advisable to check whether the valve itself is leaking before removing the tyre. Do not forget to fit the dust cap, which forms an effective extra seal.

16 Tubeless tyres: removal and refitting

1 It is strongly recommended that should a repair to a tubeless tyre be necessary, the wheel is removed from the machine and taken to a tyre fitting specialist who is willing to do the job or taken to an official dealer. This is because the force required to break the seal between the wheel rim and tyre bead is considerable and considered to be beyond the capabilities of an individual working with normal tyre removing tools. Any abortive attempt to break the rim to bead seal may also cause damage to the wheel rim, resulting in an expensive wheel replacement. If, however, a suitable bead releasing tool is available, and experience has already been gained in its use, tyre removal and refitting can be accomplished as follows.

2 Remove the wheel from the machine by following the instructions for wheel removal as described in the relevant Section of this Chapter. Deflate the tyre by removing the valve insert and when it is fully deflated, push the bead of the tyre away from the wheel rim on both sides so that the bead enters the centre well of the rim. As noted, this operation will almost certainly require the use of a bead releasing tool.

3 Insert a tyre lever close to the valve and lever the edge of the tyre over the outside of the wheel rim. Very little force should be necessary; if resistance is encountered it is probably due to the fact that the tyre beads have not entered the well of the wheel rim all the way round the tyre. Should the initial problem persist, lubrication of the tyre bead and the inside edge and lip of the rim will facilitate removal. Use a recommended lubricant, a diluted solution of washing-up liquid or french chalk. Lubrication is usually recommended as an aid to tyre fitting but is use is equally desirable during removal. The risk of lever damage to wheel rims can be minimised by the use of proprietary plastic rim protectors placed over the rim flange at the point where the tyre levers are inserted. Suitable rim protectors may be fabricated very easily from short lengths (4 – 6 inches) of thick-walled nylon petrol pipe which have been split down one side using a sharp knife. The use of rim protectors should be adopted whenever levers are used and, therefore, when the risk of damage is likely.

4 Once the tyre has been edged over the wheel rim, it is easy to work around the wheel rim so that the tyre is completely free on one side.

5 Working from the other side of the wheel, ease the other edge of the tyre over the outside of the wheel rim, which is furthest away. Continue to work around the rim until the tyre is freed completely from the rim.

6 Refer to the following Section for details relating to puncture repair and the renewal of tyres. See also the remarks relating to the tyre valves in Section 18.

7 Refitting of the tyre is virtually a reversal of removal procedure. If the tyre has a balance mark (usually a spot of coloured paint), as on the tyres fitted as original equipment, this must be positioned alongside the valve. SImilarly, any arrow indicating direction of rotation must face the right way.

8 Starting at the point furthest from the valve, push the tyre bead over the edge of the wheel rim until it is located in the central well. Continue to work around the tyre in this fashion until the whole of one side of the tyre is on the rim. It may be necessary to use a tyre lever during the final

Tyre changing sequence — tubed tyres

 A Deflate tyre. After pushing tyre beads away from rim flanges push tyre bead into well of rim at point opposite valve. Insert tyre lever adjacent to valve and work bead over edge of rim.

 B Use two levers to work bead over edge of rim. Note use of rim protectors.

 C Remove inner tube from tyre.

 D When first bead is clear, remove tyre as shown.

 E When fitting, partially inflate inner tube and insert in tyre.

 F Work first bead over rim and feed valve through hole in rim. Partially screw on retaining nut to hold valve in place.

 G Check that inner tube is positioned correctly and work second bead over rim using tyre levers. Start at a point opposite valve.

 H Work final area of bead over rim whilst pushing valve inwards to ensure that inner tube is not trapped.

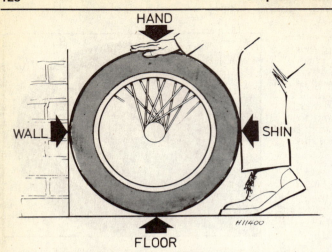

Fig. 6.14 Method of seating the beads on tubeless tyres

stages. Here again, the use of a lubricant will aid fitting. It is recommended strongly that when refitting the tyre only a recommended lubricant is used because such lubricants also have sealing properties. Do not be over generous in the application of lubricant or tyre creep may occur.

9　Fitting the upper bead is similar to fitting the lower bead. Start by pushing the bead over the rim and into the well at a point diametrically opposite the tyre valve. Continue working round the tyre, each side of the starting point, ensuring that the bead opposite the working area is always in the well. Apply lubricant as necessary. Avoid using tyre levers unless absolutely essential, to help reduce damage to the soft wheel rim. The use of the levers should be required only when the final portion of bead is to be pushed over the rim.

10　Lubricate the tyre beads again prior to inflating the tyre, and check that the wheel rim is evenly positioned in relation to the tyre beads. Inflation of the tyre may well prove impossible without the use of a high pressure air hose. The tyre will retain air completely only when the beads are firmly against the rim edges at all points and it may be found when using a foot pump that air escapes at the same rate as it is pumped in. This problem may also be encountered when using an air hose on new tyres which have been compressed in storage and by virtue of their profile hold the beads away from the rim edges. To overcome this difficulty, a tourniquet may be placed around the circumference of the tyre, over the central area of the tread. The compression of the tread in this area will cause the beads to be pushed outwards in the desired direction. The type of tourniquet most widely used consists of a length of hose closed at both ends with a suitable clamp fitted to enable both ends to be connected. An ordinary tyre valve is fitted at one end of the tube so that after the hose has been secured around the tyre it may be inflated, giving a constricting effect. Another possible method of seating beads to obtain initial inflation is to press the tyre into the angle between a wall and the floor. With the airline attached to the valve additional pressure is then applied to the tyre by the hand and shin, as shown in the accompanying illustration. The application of pressure at four points around the tyre's circumference whilst simultaneously applying the airhose will often effect an initial seal between the tyre beads and wheel rim, thus allowing inflation to occur.

11　Having successfully accomplished inflation, increase the pressure to 40 psi and check that the tyre is evenly disposed on the wheel rim. This may be judged by checking that the thin positioning line found on each tyre wall is equidistant from the rim around the total circumference of the tyre. if this is not the case, deflate the tyre, apply additional lubrication and reinflate. Minor adjustments to the tyre position may be made by bouncing the wheel on the ground.

12　Always run the tyre at the recommended pressures and never under- or over-inflate. The correct pressures for various weights and configurations are given in Routine maintenance.

17　Tubeless tyres: puncture repair and tyre renewal

1　The primary advantage of the tubeless tyre is its ability to accept penetration by sharp objects such as nails etc without loss of air. Even if loss of air is experienced, because there is no inner tube to rupture, in normal conditions a sudden blow-out is avoided. If a puncture of the tyre occurs, the tyre should be removed for inspection for damage before any attempt is made at remedial action. The temporary repair of a punctured tyre by inserting a plug from the outside should not be attempted. Although this type of temporary repair is used widely on cars, the manufacturers strongly recommend that no such repair is carried out on a motorcycle tyre. Not only does the tyre have a thinner carcass, which does not give sufficient support to the plug, but the consequences of a sudden deflation are often sufficiently serious that the risk of such an occurrence should be avoided at all costs.

2　The tyre should be inspected both inside and out for damage to the carcass. Unfortunately the inner lining of the tyre – which takes the place of the inner tube – may easily obscure any damage and some experience is required in making a correct assessment of the tyre condition.

3　There are two main types of tyre repair which are considered safe for adoption in repairing tubeless motorcycle tyres. The first type of repair consists of inserting a mushroom-headed plug into the hole from the inside of the tyre. The hole is prepared for insertion of the plug by reaming and the application of an adhesive. The second repair is carried out by buffing the inner lining in the damaged area and applying a cold or vulcanised patch. Because both inspection and repair, if they are to be carried out safely, require experience in this type of work, it is recommended that the tyre be placed in the hands of a repairer with the necessary skills, rather than repaired in the home workshop.

4　In the event of an emergency, the only recommended 'get-you-home' repair is to fit a standard inner tube of the correct size. If this course of action is adopted, care should be taken to ensure that the cause of the puncture has been removed before the inner tube is fitted. It will be found that the valve hole in the rim is considerably larger than the diameter of the inner tube valve stem. To prevent the ingress of road dirt, and to help support the valve, a spacer should be fitted over the valve.

5　In the event of the unavailability of tubeless tyres, ordinary tubed tyres fitted with inner tubes of the correct size may be fitted. Refer to the manufacturer or a tyre fitting specialist to ensure that only a tyre and tube of equivalent type and suitability is fitted, and also to advise on the fitting of a valve nut to the rim hole.

18　Tyre valves: general

Tubed tyres

1　Valve cores seldom give trouble, but do not last indefinitely. Dirt under the seating will cause a puzzling 'slow-puncture'. Check that they are not leaking by applying spittle to the end of the valve and watching for air bubbles.

2　A valve cap is a safety device, and should always be fitted. Apart from keeping dirt out of the valve, it provides a second seal in case of valve failure, and may prevent an accident resulting from sudden deflation.

Tubeless tyres

3　It will be appreciated from the preceding Sections that the adoption of tubeless tyres had made it necessary to modify the valve arrangement, as there is no longer an inner tube which can carry the valve core. The problem has been overcome by fitting a separate rubber-bodied tyre valve which passes through a close-fitting hole in the rim. The valve is fitted from the rim well, and it follows that the valve can be removed and replaced only when the tyre has been removed from the rim. Leakage of air from around the valve body is likely to occur only if the sealing seat fails.

4　The valve core is of the same types that used with tubed tyres, and screws into the valve body. The core can be removed with a small slotted tool which is normally incorporated in plunger type pressure gauges. Some valve dust caps incorporate a projection for removing valve cores. Although tubeless tyre valves seldom give trouble, it is possible for a leak to develop if a small particle of grit lodges on the sealing face. Occasionally, an elusive slow-puncture can be traced to a leaking valve core, and this should be checked before a genuine puncture is suspected.

Tyre changing sequence — tubeless tyres

Deflate tyre. After releasing beads, push tyre bead into well of rim at point opposite valve. Insert lever adjacent to valve and work bead over edge or rim.

Use two levers to work bead over edge of rim. Note use of rim protectors.

When first bead is clear, remove tyre as shown.

Before fitting, ensure that tyre is suitable for wheel. Take note of any sidewall markings such as direction of rotation arrows.

Work first bead over the rim flange.

Use a tyre lever to work the second bead over rim flange.

Chapter 6 Wheels, brakes and tyres

5   The valve dust caps are a significant part of the tyre valve assembly. Not only do they prevent the ingress of road dirt in the valve, but also act as a secondary seal which will reduce the risk of sudden deflation if a valve core should fail.

19 Wheel balancing

1   It is customary to balance the wheels complete with the tyre and, where fitted, the tube. The out-of-balance forces which exist are eliminated and the handling of the machine is improved in consequence A wheel which is badly out of balance produces through the steering a most unpleasant hammering effect at high speeds.

2   Some tyres have a balance mark on the sidewall, usually in the form of a coloured spot. This mark must be in line with the tyre valve, when the tyre is fitted to the inner tube. Even then the wheel may require the addition of balance weights, to offset the weight of the tyre valve itself.

3   If the wheel is raised clear of the ground and is spin, it will probably come to rest with the tyre valve or the heaviest part downward and will always come to rest in the same position. Balance weights must be added to a point diametrically opposite this heavy spot until the wheel will come to rest in *any* position after it is spin.

4   Note that it may be necessary to remove the caliper so that the wheel is completely free to rotate without brake drag. Special weights are available for TZR wheels from Yamaha dealers in 10, 20 and 30 gram sizes; these weights being constructed to clamp on to each side of the raised rib cast on the wheel rim. While balance weights are not specifically listed by Yamaha for the DT model, weights of a suitable type will be available at any good motorcycle dealer or tyre fitting agency. The usual type is designed to be clamped around a spoke, next to the rim, and will be available in many different sizes. Finally, note that it will be rarely necessary to balance the rear wheel but that the procedure is exactly the same as for the front wheel. The only difference is that the drive chain must be removed from the sprocket to allow the wheel to rotate freely.

Chapter 7 Electrical system

Contents

General description.. 1	Switches: examination and testing.. 8
Electrical system: general information and preliminary checks 2	Fuse: location and renewal.. 9
Battery: examination and maintenance... 3	Horn: location and testing.. 10
Charging system: checking the output... 4	Turn signal relay: location and testing..................................... 11
Lighting system: checking the output – DT models only................ 5	Low oil level warning lamp circuit: testing.............................. 12
Generator coils: testing... 6	Coolant temperature gauge circuit: testing............................. 13
Regulator/rectifier unit: location and testing................................. 7	Bulbs: renewal... 14

Specifications

Battery

	TZR model	DT model
Type...	12N5-3B	GM3-3B/FB3L-B
Capacity..	12V 5AH	12V 3AH
Earth...	Negative	Negative
Specific gravity.................................	1.280	1.280

Generator

	TZR model	DT model
Charging system voltage at 3000 rpm	14.3 – 15.3 volts	13.3 – 15.3 volts
Lighting system voltage at 2500 rpm	Not applicable	11.5 – 13.0 volts
Charging coil resistance @ 20°C (68°F).................	0.3 – 0.4 ohms	0.3 – 0.5 ohms
Lighting coil resistance @ 20°C (68°F)..................	Not applicable	0.24 – 0.36 ohms

Fuse

	TZR model	DT model
	20A	10A

Bulbs

Headlamp...	12V, 45/40W
Parking lamp.....................................	12V, 4W
Stop/tail lamp....................................	12V, 5/21W
Turn signal lamps..............................	12V, 21W
Instrument lamps..............................	12V, 3.4W
All warning lamps.............................	12V, 3W

Torque settings

Component	kgf m	lbf ft
Temperature sender unit....................	1.5	11
Generator rotor nut............................	8.0	58
Stator plate retaining screws.............	0.8	5.8

1 General description

On TZR models the power for the complete electrical system is provided by a three-phase generator mounted on the crankshaft left-hand end, the output from which is rectified and controlled by a combined regulator/rectifier unit before being passed to the battery and thence to the main electrical circuit.

The DT models employ a much simpler system, powered by a generator mounted on the crankshaft left-hand end. The three coils mounted on the generator stator plate independently power the charging, lighting and ignition system. A combined regulator/rectifier is fitted. The regulator side of the unit controls the current being fed to the bulbs to prevent them from blowing through voltage surges, and soaks up the excess current generated when the lights are switched off. This current is dispersed in the form of heat. The rectifier converts the remainder of the generator output to direct current (dc) which is then used to charge the battery, which powers the ancillary electrical equipment such as the horn, stoplamp and turn signals.

2 Electrical system: general information and preliminary checks

1 In the event of an electrical system fault, always check the physical condition of the wiring and connectors before attempting any of the test procedures described here and in subsequent Sections. Look for chafed, trapped or broken electrical leads and repair or renew these as necessary. Leads which have broken internally are not easily spotted, but may

Chapter 7 Electrical system

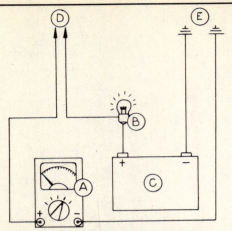

Fig. 7.1 Simple circuit testing equipment

- A Multimeter
- B Bulb
- C Battery
- D Positive probe
- E Negative probe

be checked using a multimeter or a simple battery and bulb circuit as a continuity tester. This arrangement is shown in the accompanying illustration. The various multi-pin connectors are generally trouble-free but may corrode if exposed to water. Clean them carefully, scraping off any surface deposits, and pack with silicone grease during assembly to avoid recurrent problems. The same technique can be applied to the handlebar switches.

2 The wiring harness is colour-coded and will correspond with the wiring diagram at the end of this Manual. When socket connections are used, they are designed so that reconnection can be made only in the correct position.

3 Visual inspection will usually show whether there are any breaks or frayed outer coverings which will give rise to short circuits. Occasionally a wire may become trapped between two components, breaking the inner core but leaving the more resilient outer cover intact. This can give rise to mysterious intermittent or total circuit failure. Another source of trouble may be the snap connectors and sockets, where the connector has not been pushed fully home in the outer housing, or where corrosion has occurred.

4 Intermittent short circuits can often be traced to a chafed wire that passes through or is close to a metal component such as a frame member. Avoid tight bends in the lead or situations where a lead can become trapped between casings.

5 A sound, fully charged battery, is essential to the normal operation of the system. There is no point in attempting to locate a fault if the battery is partly discharged or worn out. Check battery condition and recharge or renew the battery before proceeding further.

6 Many of the test procedures described in this Chapter require voltages or resistances to be checked. This necessitates the use of some form of test equipment such as a simple and inexpensive multimeter of the type sold by electronics or motor accessory shops.

7 If you doubt your ability to check the electrical system, entrust the work to an authorised Yamaha dealer. In any event have your findings double-checked before consigning expensive components to the scrapbin.

3 Battery: examination and maintenance

1 Details of the regular checks needed to maintain the battery in good condition are given in Routine maintenance, together with instructions on removal and refitting and general battery care. Batteries can be dangerous if mishandled; read carefully the 'Safety first' section at the front of this Manual before starting work, and always wear overalls or old clothing in case of accidental acid spillage. If acid is ever allowed to splash into your eyes or on to your skin, flush it away with copious quantities of fresh water and seek medical advice immediately.

2 When new, the battery is filled with an electrolyte of dilute sulphuric acid having a specific gravity of 1.280 at 20°C (68°F). Subsequent evaporation, which occurs in normal use, can be compensated for by topping up with distilled or demineralised water only. Never use tap water as a substitute and do not add fresh electrolyte unless spillage has occurred.

3 The state of charge of a battery can be checked using an hydrometer.

4 The normal charge rate for a battery is $\frac{1}{10}$ of its rated capacity, thus for a 5 ampere hour unit charging should take place at 0.5 amp. Exceeding this figure could cause the battery to overheat, buckling the plates and rendering it useless. Few owners will have access to an expensive current controlling charger, so if a normal domestic charger is used check that after a possible initial peak, the charge rate falls to a safe level. If the battery becomes hot during charging **stop**. Further charging will cause damage. Note that the cell caps should be loosened and vents unobstructed during charging to avoid a build-up of pressure and risk of explosion.

5 After charging, top up with distilled water as required, then check the specific gravity and battery voltage. Specific gravity should be above 1.270 and a sound, fully charged battery should produce 15 – 16 volts. If the recharged battery discharges rapidly if left disconnected it is likely that an internal short caused by physical damage or sulphation has occurred. A new battery will be required. A sound item will tend to lose its charge at about 1% per day.

4 Charging system: checking the output

1 Remove the right-hand sidepanel to gain access to the battery. Note the battery must be fully charged and in good condition if the test is to be accurate.

2 Connect a dc voltmeter, set to the appropriate scale, across the battery terminals (meter positive lead to battery positive terminal (+) and meter negative lead to battery negative terminal (−)), and start the engine. Slowly increase the engine speed to 3000 rpm and note the voltage reading obtained. Stop the engine. If the reading obtained differs significantly from that specified, and the battery is known to be fully charged, the charging system is faulty.

3 If the charging system is found to be at fault, further testing should be carried out to trace the faulty component in the order shown below.

   (a) Charging coil resistance, see Section 6.
   (b) Charging system wiring and connectors, see Section 2.
   (c) Regulator/rectifier unit, see Section 7.

5 Lighting system: checking the output – DT models only

1 If the lights appear to be dim or constant bulb blowing is experienced, and the bulb filaments appear to have melted, the lighting system should be checked to ensure that the right voltage is being supplied. Note the battery must be fully charged and in good condition if the test is to be accurate.

2 Remove the headlamp unit, as described in Section 14.

3 Set a voltmeter or multimeter on the 0 – 20V ac range and connect its positive lead to the green wire terminal and its negative lead to the black wire terminal of the headlamp connector. Turn the headlamp switch on, check that the lighting dip switch is set to the low beam position and start the engine. Increase the engine speed to 2500 rpm and note the reading obtained. Stop the engine. If this reading differs significantly from the specified limits the lighting system is faulty.

4 If the lighting system is found to be at fault, further testing should be carried out to trace the faulty component in the order shown below.

   (a) Lighting coil resistance, see Section 6.
   (b) All switches, wiring and connectors, see Section 2.
   (c) Regulator/rectifier unit, see Section 7.

5 If the reading is correct and the fault persists, check that the bulbs fitted are of the correct voltage and wattage, renewing them if necessary. If the bulbs are of the correct rating, the fault must lie in the wiring or bulbholder which should be checked as described in Section 2 of this Chapter.

6 Generator coils: testing

1 Trace the generator output wiring back to the connector blocks and disconnect the generator wiring from the main wiring harness. Make the following tests on the generator side of the wiring.

TZR model
2 Identify the block connector which contains the three white leads and carry out the following tests. Using a multimeter set to the ohm x 1 scale, measure the resistance between each of the three white leads (three tests) and note the readings obtained. If either value differs greatly from those specified the stator coil assembly is at fault and should be renewed.

DT model – charging coil
3 Using an ohmmeter or multimeter set to the ohms x 1 scale, measure the resistance between the white and black wire terminals. If the reading obtained differs significantly from that specified the charging coil is at fault and should be renewed.

DT model – lighting coil
4 Using an ohmmeter or a multimeter set to the ohms x 1 scale, measure the resistance between the yellow/red and black terminals. If the reading obtained differs significantly from that specified the lighting coil is faulty and should be renewed.

All models
5 If a coil is thought to be faulty, first check that this is not due to a broken or damaged wire between the coil itself and the block connector. Pinched or broken wires can usually be repaired by the average private owner.
6 If a faulty coil is confirmed the flywheel rotor must be removed, as described in Section 14 of Chapter 1 for access to the stator plate. On DT models the charging and lighting coils are available as separate items although on TZR models the complete stator assembly must be renewed. Note that it is worth seeking the advice of an auto electrical specialist, who may be able to rewind the damaged coil.

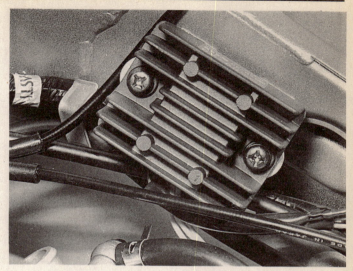

7.1a Regulator/rectifier unit location – TZR model

7 Regulator/rectifier unit: location and testing

1 The regulator/rectifier unit is a sealed, heavily finned unit mounted on the frame beneath the fuel tank. Remove the seat, both sidepanels and the fuel tank to gain access to it.
2 If the regulator/rectifier unit is thought to be faulty, it must be removed and taken to an authorized Yamaha dealer for testing. Whilst it is normally possible to test the unit by making resistance measurements across the various terminals, no such test data is available on these models. Therefore it will be necessary to enlist the help of an authorised Yamaha dealer who will be able to check the operation of the unit by substituting a sound item.

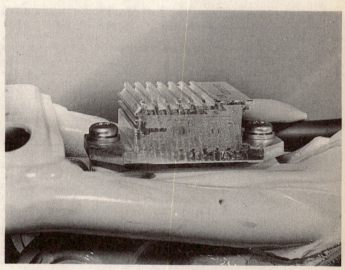

7.1b Regulator/rectifier unit location – DT model

8 Switches: examination and testing

1 The function of each switch can be easily tested using a multimeter or a battery and bulb arrangement (see Section 2) to check continuity in each of the switch positions.
2 Referring to the wiring diagram at the end of this Manual, note that each switch is represented in diagrammatic form, showing the wiring colours and their corresponding switch positions. Identify the wiring connectors belonging to the switch in question and unplug it. Connect the multimeter or battery and bulb arrangement across each pair of leads to be tested and check that it functions as shown on the switch diagram when the switch is operated. If a fault is indicated, or if operation is erratic, the switch must either be repaired or renewed as described below.

Ignition and engine kill switches
3 The ignition and engine kill switches are covered in Chapter 4, Section 7.

Handlebar switches
4 If the operation of a switch has been affected by corrosion or water, try soaking it in WD40 and operating the switch repeatedly. This is quite often successful, the back of each switch being so designed that the fluid will penetrate readily. If this fails to effect a cure, the switch will have to be renewed. Before purchasing a new switch, however, you may wish to attempt to dismantle and physically clean the switch terminals. The method of doing so is self-evident, but be warned that the switch may tend to fly apart when disturbed! It is suggested that the switch is removed and the work carried out on a clean bench so that small parts such as springs are not mislaid.

Stop lamp switches
5 The front stop lamp switch is a small plunger type unit mounted in the brake lever stock. The switch is visible at the front of the brake lever, but if removal is necessary, or access to the wiring connectors is required, the headlamp unit must first be removed as described in Section 14. If the operation of a switch has been affected by corrosion or water, try soaking it in WD40 and operating the switch repeatedly. If this fails to restore operation, renew the switch; no dismantling is possible.
6 To remove the switch, use a pointed instrument to press in its locking tab, then withdraw the switch from the lever. On refitting, carefully press the switch into place until the tab clicks into its slot.
7 The rear stop lamp switch is mounted on the right-hand side of the frame, just above the brake pedal. The switch is similar in operation to

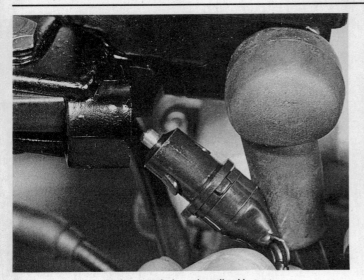
8.6 Remove front stop lamp switch as described in text

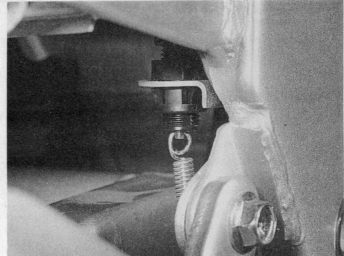
8.7 Rear stop lamp switch location – TZR125 shown

the front stop lamp switch and the remarks made above can be applied. Details of switch height adjustment can be found in Routine maintenance.

8 If access to the wiring connectors is required note that it will be necessary to remove the seat.

Neutral and sidestand switches

9 The neutral and sidestand switches are covered in Chapter 4, Section 9.

9 Fuse: location and renewal

1 The electrical system is protected by a single fuse which is retained in a plastic casing set in the battery positive (+) terminal lead, and is clipped to a holder next to the battery. if the spare fuse is ever used, replace it with one of the correct rating as soon as possible.

2 Before renewing a fuse that has blown, check that no obvious short circuit has occurred, otherwise the replacement fuse will blow immediately it is inserted. It is always wise to check the electrical circuit thoroughly, to trace the fault and eliminate it.

3 When a fuse blows while the machine is running and no spare is available, a 'get you home' remedy is to remove the blown fuse and wrap it in silver paper before replacing it in the fuse holder. The silver paper will restore the electrical continuity by bridging the broken fuse wire. This expedient should never be used if there is evidence of short circuit or other major electrical faults, otherwise more serious damage will be caused. Replace the 'doctored' fuse at the earliest possible opportunity, to restore full circuit protection.

10 Horn: location and testing

1 The horn is mounted on the bottom yoke of the TZR models, and is bolted to the frame, just below the radiator, on DT models. No maintenance is required other than regular cleaning to remove road dirt and occasional spraying with WD40 or a similar water dispersant spray to minimise internal corrosion.

2 If the horn fails to work, first check that the battery is fully charged. If full power is available, a simple test will reveal whether the current is reaching the horn. Disconnect the horn wires and substitute a 12 volt bulb. Switch on the ignition and press the horn button. If the bulb fails to light, check the horn button and wiring as described in Sections 2 and 8 of this Chapter. If the bulb does light, the horn circuit is proved good and the horn itself must be checked.

3 With the horn wires still disconnected, connect a fully charged 12 volt battery directly to the horn. If it does not sound, a sharp tap on the outside may serve to free the internal contacts. If this fails, the horn must be renewed as repairs are not possible.

4 Different types of horn may be fitted; if a screw and locknut is provided on the outside of the horn, the internal contacts may be adjusted to compensate for wear and to cure a weak or intermittent horn note. Slacken the locknut and rotate slowly the screw until the clearest and loudest note is obtained, then retighten the locknut. If no means of adjustment is provided on the horn fitted, it must be renewed.

11 Turn signal relay: location and testing

1 The turn signal relay is a sealed unit black plastic unit, rubber-mounted to protect it from vibration. On TZR models it is mounted on the rear mudguard, between the oil tank and expansion tank filler caps and it will be necessary to remove the seat to gain access to it. On DT models the turn signal relay is located behind the headlamp unit, necessitating removal of the headlamp fairing and unit, as described in Section 14, before it can be reached.

2 If the turn signal lamps cease to function correctly, there may be any one of several possible faults responsible which should be checked before the relay is suspected. First check that the lamps are correctly mounted and that all the earth connections are clean and tight. Check that the bulbs are of the correct wattage and that corrosion has not

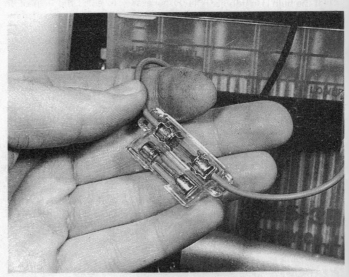
9.1 Fuse location – always carry a spare fuse of the correct rating

Chapter 7 Electrical system

10.1 Horn location – TZR model

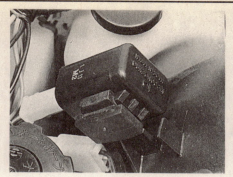

11.1a The turn signal relay is situated between oil and expansion tanks on TZR model ...

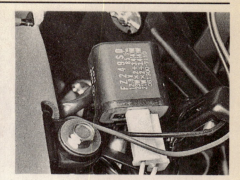

11.1b ... and behind headlamp unit on DT model

developed on the bulbs or in their holders. Any such corrosion must be thoroughly cleaned off to ensure proper bulb contact. Also check that the turn signal switch is functioning correctly and that the wiring is in good order. Finally ensure that the battery is fully charged.

3 Faults in any one or more of the above items will produce symptoms for which the turn signal relay may be blamed unfairly. If the fault persists even after the preliminary checks have been made, the relay must be at fault. Unfortunately the only practical method of testing the relay is to substitute a known good one. If the fault is then cured, the relay is proven faulty and must be renewed.

12 Low oil level warning lamp circuit: testing

1 The circuit consists of a sender unit, mounted in the top of the oil tank, and the bulb itself, which is mounted in the instrument panel (TZR model) or in the tachometer (DT model).

2 With the ignition switched on, and the transmission in neutral, the oil level warning lamp should light up as a check that the system is working. The lamp should go out as soon as the transmission is put into gear. If not, there is insufficient oil in the tank; top up the tank as described in Routine maintenance and check again that the lamp goes out. If the lamp still does not go out or does not come on at all the system is faulty and must be checked.

3 If the bulb fails to light at all, the most likely cause of failure will be a blown bulb, but it is worth checking first that the battery is fully charged; check that all other systems are working normally as a quick test of this. The bulb is removed and refitted as described in Section 14 of this Chapter and Section 17 of Chapter 5.

4 If the bulb is in good condition the fault must lie in the switch or in the wiring. Check the relevant wiring, as described in Section 2 of this Chapter, using the wiring diagram at the end of this Chapter to trace the circuit. If the bulb and wiring are in good condition the fault must lie in the tank-mounted sender unit.

5 Remove the oil level warning sender unit, as described in Section 13 of Chapter 3 and test it as follows. Using a multimeter set to the ohms x 1 scale, check for continuity between the black/red and light blue wire terminals. If the switch is functioning correctly there should be continuity between the terminals. Also check for continuity between the black/red and black wire terminals. If the switch is functioning correctly there should be continuity between the terminals when the sender unit is held in an upright position and an open circuit when the sender unit is held upside down. If either or both tests do not produce the expected results the sender unit should be renewed.

13 Coolant temperature gauge circuit: testing

1 The circuit consists of the sender unit mounted in the cylinder head water jacket and the gauge assembly mounted in the instrument panel (TZR model) or tachometer (DT model).

2 To test the system, first ensure that the battery is fully charged by checking that all other electrical circuits work properly, then disconnect its wire from the sender unit and check that the ignition is switched on. The temperature gauge needle should point to 'C'. Earth the sender unit wire on the cylinder head, whereupon the needle should immediately swing over to 'H'. Do not earth the wire for more than 5 seconds or the gauge may be damaged.

3 If the needle moves as described above, a fault in the temperature sender unit is indicated and it should be removed from the head for further testing as described below. However, if the needle fails to move, the gauge or wiring is faulty and should be examined further; proceed to paragraph 8.

4 To remove the sender unit disconnect the wire leading to it, then slacken it using a suitable spanner. Unscrew the unit as fast as possible and withdraw it, swiftly plugging the opening to stop the coolant escaping. Refer to paragraph 5 for testing details. On refitting check the condition of the sealing washer renewing it if necessary, and screw the sender unit into place; do not overtighten – note the torque setting given in the Specifications. If sufficient care is taken, the minimum of coolant will escape and there will be no need to drain and refill the system, but remember to top up the radiator or expansion tank, and to wash off all the spilt coolant. Note that the cooling system must be bled of air as described in Chapter 2, Section 4.

5 If the coolant temperature sender appears to be faulty it can be tested by measuring its resistance at various temperatures. To accomplish this it will be necessary to gather together a heatproof container into which the sender can be placed, a burner of some description (a small gas-powered camping burner would be ideal), a thermometer capable of measuring between 40°C and 120°C (122°F – 248°F) and an ohmmeter or multimeter capable of measuring 0 – 200 ohms with a reasonable degree of accuracy.

6 Fill the container with cold water and arrange the sender unit on some wire so that the probe end is immersed in it. Connect one of the meter leads to the sender body and the other to the terminal. Suspend the thermometer so that the bulb is close to the sender probe.

7 Start to heat the water, and make a note of the resistance reading at

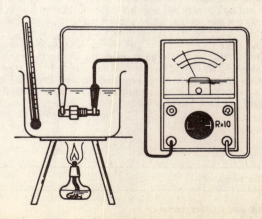

Fig. 7.2 Coolant temperature sender unit test

14.2a Release headlamp unit by removing the bolts (DT model) ...

14.2b ... or screws (TZR model) ....

14.2c ... and unplugging wiring connector

14.2d Parking lamp bulbholder is a bayonet fit in headlamp unit

14.3a Remove rubber cover ...

14.3b ... then bulbholder ring ...

the temperature shown in the table below. If the unit does not give readings which approximate quite closely to those shown it must be renewed.

Coolant temperature 50°C (122°F)   80°C (176°F)   100°C (212°F)
Sender resistance   153.9Ω   47.5 – 56.8Ω   26.2 – 29.3Ω

8   If the gauge appears to be faulty, dismantle the instrument panel, as described in Section 17 of Chapter 5, and check the relevant wiring and connectors as described in Section 2 of this Chapter. If all appears to be well it will be necessary to check the power supply from the ignition switch.

9   Disconnect the switch from the main wiring harness at its block connector, and using a ohmmeter or multimeter set to the ohms x 1 scale, check for continuity between the red and brown terminals on the switch side of the wiring. If the switch is in a good condition there should be continuity between the terminals when the switch is in the ON position and an open circuit (infinite resistance) when it is in the OFF position. If the results are not as shown the ignition switch must be renewed.

10 If the tests carried out on the sender unit, ignition switch and wiring show that they are in good condition, the fault must be in the gauge assembly; this must be renewed as repairs are not possible. On TZR models the gauge can be renewed individually, but on DT models it forms a part of the sealed tachometer unit and the complete unit must be renewed to cure a gauge fault.

14 Bulbs: renewal

Headlamp and parking lamp

1   On DT models it will first be necessary to remove the headlamp fairing to gain access to the headlamp unit.

2   Remove both the headlamp retaining bolts (DT models) or screws (TZR models) and partially withdraw the headlamp assembly until the wiring can be reached. Unplug the headlamp bulb wiring connector, then remove the parking lamp bulbholder from the headlamp by pressing it in and turning it anticlockwise. The headlamp unit can then be lifted clear.

3   Remove the rubber bulbholder cover and release the holder by pressing it in and turning it anticlockwise. The headlamp bulb can then

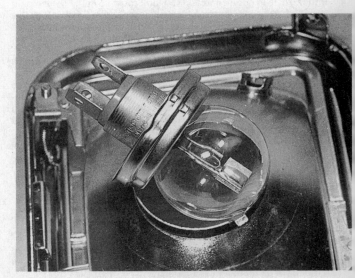
14.3c ... and withdraw headlamp bulb

Chapter 7 Electrical system

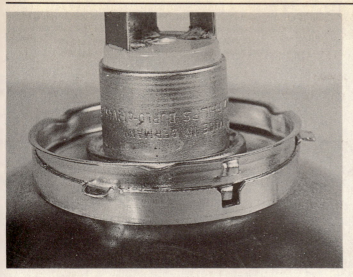

14.3d Ensure tags on bulb locate in slot in headlamp reflector

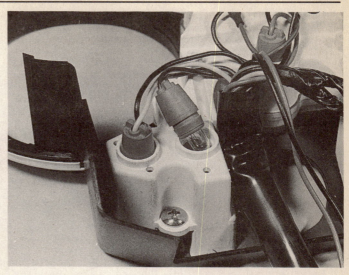

14.4 Instrument and warning lamp bulbholders are a press fit in panel assembly

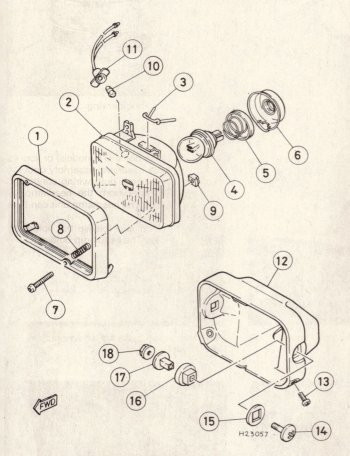

Fig. 7.3 Headlamp – TZR models

1 Rim
2 Reflector unit
3 Retaining spring
4 Headlamp bulb
5 Bulbholder
6 Rubber cover
7 Beam adjusting screw – 2 off
8 Spring – 2 off
9 Nut – 2 off
10 Parking lamp bulb
11 Bulbholder
12 Headlamp shell
13 Screw – 2 off
14 Screw – 2 off
15 Damping rubber – 2 off
16 Damping rubber – 2 off
17 Spacer – 2 off
18 Nut – 2 off

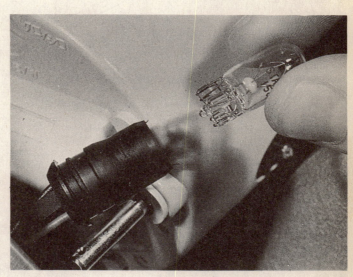

14.5 Bulbs are of the capless type

be lifted out of the unit. The parking lamp bulb should be pressed in and twisted anticlockwise to release it from its holder. On refitting make sure the tabs on the headlamp bulb are correctly positioned in the reflector. Check the headlamp beam settings, as described in Routine maintenance.

Instrument illuminating and warning lamps
4 All are fitted into bulbholders which are pressed into the base of the instrument panel. Refer to Chapter 5, Section 17 for details of the work necessary to reach them.
5 Unplug the bulbholder concerned, then remove the bulb. All the bulbs are of the capless type which simply pull out of the holder. On refitting, take care not to damage the fine wire tails of the bulb, and ensure that they are pushed fully into their contacts.

Turn signals and stop/tail lamp
6 With the exception of the tail lamp on TZR models, all lamp lenses are retained by screws. Remove the retaining screws which secure the lamp lens and withdraw the lens. In the case of the TZR model tail lamp, carefully prise the lens out of its retaining lugs and lift it clear. Note that it might be necessary to first remove the small panel from beneath the

grab rail to gain access to these lugs. On all lenses make sure that the seal between the lens and body is not damaged, renewing it if necessary. The bulb should be pressed in and rotated anticlockwise to release it.

7 Refitting is the reverse of the removal procedure. Note that on the stop/tail lamp bulb the locating pins are offset so that the bulb can only be refitted the correct way. Do not overtighten the lens screws or either the lens will be cracked or the plastic body will be damaged.

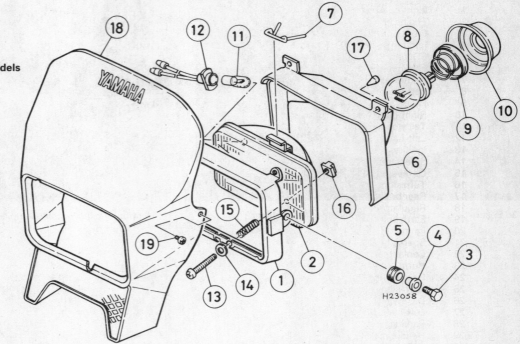

Fig. 7.4 Headlamp – DT models

1 Rim
2 Reflector unit
3 Bolt – 2 off
4 Spacer – 2 off
5 Grommet – 2 off
6 Headlamp shell
7 Retaining spring
8 Headlamp bulb
9 Bulbholder
10 Rubber cover
11 Parking lamp bulb
12 Bulbholder
13 Beam adjusting screw – 2 off
14 Washer – 2 off
15 Spring – 2 off
16 Nut – 2 off
17 Clip – 4 off
18 Headlamp fairing
19 Screw – 2 off

14.6a Turn signal lamp lenses are retained by two screws on TZR model ...

14.6b ... and a single screw on DT model

14.6c Turn signal bulbs are a bayonet fit in their holders

14.6d The DT model lamp lens is retained by two screws and uses a single bulb ...

14.6e ... whereas the TZR model lens is retained by lugs (arrowed) ...

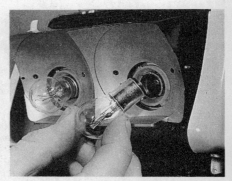

14.6f ... and uses a twin bulb arrangement. Note offset pins on bulb

Component key – TZR models

1 Lighting switch
2 Engine kill switch
3 Front brake lamp switch
4 Ignition switch
5 CDI unit
6 Flywheel generator
7 Neutral switch
8 Battery
9 Fuse
10 Right-hand rear turn signal
11 Tail/stop lamps
12 Left-hand rear turn signal
13 Ignition control unit
14 Sidestand switch
15 Oil level sender unit
16 Turn signal relay
17 Rear brake lamp switch
18 Regulator/rectifier unit
19 Earth point
20 Ignition HT coil
21 Spark plug
22 Coolant temperature sender unit
23 Horn
24 Horn switch
25 Turn signal switch
26 Dip switch
27 Left-hand front turn signal
28 Parking lamp
29 Headlamp
30 Oil level warning lamp
31 High beam lamp
32 Instrument lamps
33 Turn signal warning lamp
34 Neutral lamp
35 Coolant temperature gauge
36 Right-hand front turn signal

Component key - DT models

1 Horn
2 Ignition HT coil
3 Spark plug
4 Earth point
5 CDI unit
6 Flywheel generator
7 Regulator/rectifier unit
8 Neutral switch
9 Ignition control unit
10 Battery
11 Fuse
12 Right-hand rear turn signal
13 Left-hand rear turn signal
14 Tail/stop lamp
15 Sidestand switch
16 Rear brake lamp switch
17 Turn signal relay
18 Coolant temperature sender unit
19 Oil level sender unit
20 Dip switch
21 Turn signal switch
22 Horn switch
23 Lighting switch
24 Ignition switch
25 Left-hand front turn signal
26 Parking lamp
27 Headlamp
28 High beam lamp
29 Oil level warning lamp
30 Turn signal warning lamp
31 Neutral lamp
32 Instrument lamps
33 Coolant temperature gauge
34 Engine kill switch
35 Front brake lamp switch
36 Right-hand front turn signal

Wire colour key

B Black
Br Brown
Ch Dark brown
Dg Dark green
G Green
L Blue
O Orange
P Pink
R Red
Sb Light blue
W White
Y Yellow

Wiring diagram – TZR 2RK and 3PC1 models
See page 139 for key

141

Wiring diagram – TZR 3PC2 and 3PC3 models
See page 139 for key

Wiring diagram – DT 3DB1 model
See page 139 for key

Wiring diagram – DT 3RN1 models onward
See page 139 for key

Conversion factors

Length (distance)
Inches (in)	X	25.4	= Millimetres (mm)	X 0.0394	= Inches (in)
Feet (ft)	X	0.305	= Metres (m)	X 3.281	= Feet (ft)
Miles	X	1.609	= Kilometres (km)	X 0.621	= Miles

Volume (capacity)
Cubic inches (cu in; in³)	X	16.387	= Cubic centimetres (cc; cm³)	X 0.061	= Cubic inches (cu in; in³)
Imperial pints (Imp pt)	X	0.568	= Litres (l)	X 1.76	= Imperial pints (Imp pt)
Imperial quarts (Imp qt)	X	1.137	= Litres (l)	X 0.88	= Imperial quarts (Imp qt)
Imperial quarts (Imp qt)	X	1.201	= US quarts (US qt)	X 0.833	= Imperial quarts (Imp qt)
US quarts (US qt)	X	0.946	= Litres (l)	X 1.057	= US quarts (US qt)
Imperial gallons (Imp gal)	X	4.546	= Litres (l)	X 0.22	= Imperial gallons (Imp gal)
Imperial gallons (Imp gal)	X	1.201	= US gallons (US gal)	X 0.833	= Imperial gallons (Imp gal)
US gallons (US gal)	X	3.785	= Litres (l)	X 0.264	= US gallons (US gal)

Mass (weight)
Ounces (oz)	X	28.35	= Grams (g)	X 0.035	= Ounces (oz)
Pounds (lb)	X	0.454	= Kilograms (kg)	X 2.205	= Pounds (lb)

Force
Ounces-force (ozf; oz)	X	0.278	= Newtons (N)	X 3.6	= Ounces-force (ozf; oz)
Pounds-force (lbf; lb)	X	4.448	= Newtons (N)	X 0.225	= Pounds-force (lbf; lb)
Newtons (N)	X	0.1	= Kilograms-force (kgf; kg)	X 9.81	= Newtons (N)

Pressure
Pounds-force per square inch (psi; lbf/in²; lb/in²)	X	0.070	= Kilograms-force per square centimetre (kgf/cm²; kg/cm²)	X 14.223	= Pounds-force per square inch (psi; lbf/in²; lb/in²)
Pounds-force per square inch (psi; lbf/in²; lb/in²)	X	0.068	= Atmospheres (atm)	X 14.696	= Pounds-force per square inch (psi; lbf/in²; lb/in²)
Pounds-force per square inch (psi; lbf/in²; lb/in²)	X	0.069	= Bars	X 14.5	= Pounds-force per square inch (psi; lbf/in²; lb/in²)
Pounds-force per square inch (psi; lbf/in²; lb/in²)	X	6.895	= Kilopascals (kPa)	X 0.145	= Pounds-force per square inch (psi; lbf/in²; lb/in²)
Kilopascals (kPa)	X	0.01	= Kilograms-force per square centimetre (kgf/cm²; kg/cm²)	X 98.1	= Kilopascals (kPa)
Millibar (mbar)	X	100	= Pascals (Pa)	X 0.01	= Millibar (mbar)
Millibar (mbar)	X	0.0145	= Pounds-force per square inch (psi; lbf/in²; lb/in²)	X 68.947	= Millibar (mbar)
Millibar (mbar)	X	0.75	= Millimetres of mercury (mmHg)	X 1.333	= Millibar (mbar)
Millibar (mbar)	X	0.401	= Inches of water (inH$_2$O)	X 2.491	= Millibar (mbar)
Millimetres of mercury (mmHg)	X	0.535	= Inches of water (inH$_2$O)	X 1.868	= Millimetres of mercury (mmHg)
Inches of water (inH$_2$O)	X	0.036	= Pounds-force per square inch (psi; lbf/in²; lb/in²)	X 27.68	= Inches of water (inH$_2$O)

Torque (moment of force)
Pounds-force inches (lbf in; lb in)	X	1.152	= Kilograms-force centimetre (kgf cm; kg cm)	X 0.868	= Pounds-force inches (lbf in; lb in)
Pounds-force inches (lbf in; lb in)	X	0.113	= Newton metres (Nm)	X 8.85	= Pounds-force inches (lbf in; lb in)
Pounds-force inches (lbf in; lb in)	X	0.083	= Pounds-force feet (lbf ft; lb ft)	X 12	= Pounds-force inches (lbf in; lb in)
Pounds-force feet (lbf ft; lb ft)	X	0.138	= Kilograms-force metres (kgf m; kg m)	X 7.233	= Pounds-force feet (lbf ft; lb ft)
Pounds-force feet (lbf ft; lb ft)	X	1.356	= Newton metres (Nm)	X 0.738	= Pounds-force feet (lbf ft; lb ft)
Newton metres (Nm)	X	0.102	= Kilograms-force metres (kgf m; kg m)	X 9.804	= Newton metres (Nm)

Power
Horsepower (hp)	X	745.7	= Watts (W)	X 0.0013	= Horsepower (hp)

Velocity (speed)
Miles per hour (miles/hr; mph)	X	1.609	= Kilometres per hour (km/hr; kph)	X 0.621	= Miles per hour (miles/hr; mph)

*Fuel consumption**
Miles per gallon, Imperial (mpg)	X	0.354	= Kilometres per litre (km/l)	X 2.825	= Miles per gallon, Imperial (mpg)
Miles per gallon, US (mpg)	X	0.425	= Kilometres per litre (km/l)	X 2.352	= Miles per gallon, US (mpg)

Temperature

Degrees Fahrenheit = (°C x 1.8) + 32   Degrees Celsius (Degrees Centigrade; °C) = (°F - 32) x 0.56

*It is common practice to convert from miles per gallon (mpg) to litres/100 kilometres (l/100km), where mpg (Imperial) x l/100 km = 282 and mpg (US) x l/100 km = 235

Index

A

Accessories 11
Adjustment:-
  brake 33, 34
  carburettor 30
  clutch 31, 32
  final drive chain 26 – 28
  oil pump 30, 91, 92
  rear suspension unit 109, 110
  spark plug gap 28
  stop lamp switch 33
  throttle cable 30
Air filter 28, 30, 88

B

Battery:-
  check 24, 25
  examination and maintenance 132
Balancing – wheel 130
Balancer shaft:-
  refitting 58, 59
  removal 49, 50
Balancer shaft gears:-
  examination and renovation 53
  refitting 60 – 62
  removal 48
Bearings:-
  engine 50, 51, 58
  rear suspension 35, 106 – 108
  steering head 37, 104, 105
  wheel 34, 35, 37, 118, 119
Bleeding:-
  brakes 124, 125
  oil pump 92
Brakes:-
  check 25
  fault diagnosis 20, 21
  disc brake:
    bleeding 124, 125
    caliper 38, 120, 121
    disc 124
    fluid renewal 36
    hoses 38
    master cylinder 38, 121 – 124
    pad check and renewal 32, 33
  drum brake:
    adjustment 34, 35
    examination and renovation 125, 126
    operating camshaft lubrication 37
  specifications 113
  stop lamp switch 33, 133

C

Cables:-
  clutch 31, 32
  lubrication 35
  oil pump 30
  throttle 30
Caliper – disc brake 120 – 121
Carburettor:-
  adjustment 30, 86, 87
  dismantling, examination and reassembly 83 – 86
  removal and refitting 82, 83
  settings 86
  specifications 78
CDI system and unit 94 – 96
Chain – final drive 26 – 28
Charging system:-
  coil test 133
  output test 132
Cleaning the machine 37
Clutch:-
  adjustment 31, 32
  examination and renovation 53, 54
  fault diagnosis 17, 18
  refitting 64, 65
  removal 46, 47
  specifications 40
Conversion factors 144
Cooling system:-
  check 35
  coolant level check 24
  coolant renewal 37
  draining 71, 72
  filling 72
  flushing 72
  hoses and connections 75
  radiator 73 – 75
  specifications 71
  temperature gauge 77, 135
  thermostat 75, 76
  water pump 76, 77
Crankcase covers:-
  left-hand 43, 51, 69
  right-hand 46, 51, 65
Crankcase halves:-
  examination and renovation 51
  joining 59
  separation 49
Crankshaft:-
  examination and renovation 52, 53
  refitting 58, 59
  removal 49, 50
Cush drive – rear wheel 120
Cylinder head and barrel:-
  examination and renovation 51, 52
  refitting 66 – 68
  removal 44, 45

D

Decarbonising:-
   engine 37
   exhaust 37, 38
Dimensions – model 6
Disc brake see **Front brake** or **Rear brake – disc**
Drum brake see **Rear brake – drum**

E

Electrical system:-
   battery 24, 132
   coolant temperature gauge 77, 135
   engine oil level warning sender 90, 135
   fault diagnosis 21, 22
   fuse 134
   general checks 131, 132
   generator – removal and refitting 49, 60
   generator – coils test 132, 133
   headlamp and parking lamp 136, 137
   horn 134
   instrument and warning lamps 137
   neutral switch 49, 60
   regulator/rectifier unit 133
   specifications 131
   stop/tail lamp 137, 138
   switches 133, 134
   turn signals 134, 137, 138
   wiring diagrams 139 – 143
Engine:-
   balancer shaft 49, 50, 58, 59
   balancer shaft gears 48, 53, 60 – 62
   bearings 50, 51, 58
   crankcase cover – left-hand 43, 51, 69
   crankcase cover – right-hand 46, 51, 65
   crankcase halves 49, 51, 59
   crankshaft 49, 50, 52, 53, 58, 59
   cylinder head and barrel 44, 45, 51, 52, 66 – 68
   decarbonising 37
   dismantling – preliminaries 44
   examination and renovation – general 50, 51
   fault diagnosis 15 – 19
   oil level check 24
   oil pump 30, 90 – 92
   oil seals 50, 51, 58
   piston and rings 44, 45, 52, 66 – 68
   primary drive gear 46, 47, 53, 60 – 62
   reassembly – general 57, 58
   reed valve 49, 59, 60, 87
   refitting in frame 68, 69
   removal from frame 42 – 44
   specifications 39 – 41
   starting and running the rebuilt engine 69
   taking the rebuilt machine on the road 69, 70
   YPVS components 46, 66
Exhaust:-
   decarbonising 37
   general 88, 89
   refitting 69
   removal 42

F

Fairing 112
Fault diagnosis 14 – 22
Filter:-
   air 28, 30, 88
   fuel 37, 82
Final drive chain 26 – 28
Footrests 110
Frame 105

Front brake:-
   caliper 38, 120 – 121
   fluid renewal 36
   lever adjustment 34
   master cylinder 38, 121 – 123
   pad check and renewal 32, 33
Front forks:-
   oil change 36
   dismantling and reassembly 99 – 102
   examination and renovation 102, 103
   refitting 103
   removal 99
   specifications 98
Front wheel:-
   bearings 34, 35, 37, 118, 119
   refitting 115, 116
   removal 114, 115
Fuel system:-
   carburettor 30, 82 – 87
   fault diagnosis 15 – 17
   fuel filters 37, 82
   fuel pipes 30, 81, 82
   fuel tank 79 – 81
   fuel tap 81, 82
   specifications 78, 79
Fuse 134

G

Gear selector components:-
   examination and renovation 54, 55
   external components:
      refitting 60
      removal 48
   internal components:
      refitting 58, 59
      removal 49, 50
Gearbox:-
   fault diagnosis 18, 19
   oil change 36, 92
   oil level check 31, 92
   shafts:
      dismantling and reassembly 55 – 57
      refitting 58, 59
      removal 49, 50
   specifications 40, 41
   sprocket:
      refitting 68
      removal 43
Gearchange lever:-
   refitting 69
   removal 43
Generator:-
   refitting 60
   removal 48, 49
   testing 94, 95, 132, 133

H

Handlebars – removal and refitting:-
   DT model 104, 105
   TZR model 99, 103
Handlebar switches 133
Headlamp:-
   beam adjustment 25
   bulb renewal 136, 137
Horn 134
Hoses – coolant 75
Hoses – hydraulic 38
HT coil 94
HT lead 97

Index

I

Ignition system:-
  CDI system 94
  CDI unit 96
  fault diagnosis 15 – 17, 94
  generator 49, 60, 94, 95
  HT coil 94
  HT lead and suppressor cap 97
  ignition control unit 96
  ignition timing 97
  pulser coil 95
  source coil 94, 95
  spark plug 28, 29, 37, 97
  specifications 93
  switches 95
Instrument:-
  drives:
    speedometer 37, 112
    tachometer 48, 62
  drive cables 35, 112
  heads 110
  lamps 137

K

Kickstart lever:-
  refitting 69
  removal 42
Kickstart assembly:-
  examination and renovation 54
  refitting 63, 64
  removal 47, 48

L

Lamps 131, 136 – 138
Lubrication:-
  cables 35
  engine:
    oil level check 24
    oil pipes 30
    oil pump 30, 90 – 92
    oil tank 89, 90
  drum brake operating camshaft 37
  final drive chain 26 – 28
  rear suspension linkage 35
  speedometer drive 37
  specifications 23, 78
  steering head bearings 37
  transmission:
    general 92
    oil change 36
    oil level check 31
  wheel bearings 37

M

Maintenance – routine 23 – 38
Master cylinder:-
  front 38, 121 – 123
  rear 38, 123, 124

N

Neutral switch:-
  refitting 60
  removal 49
  testing 110, 134

O

Oil – engine:-
  level check 24
  level warning sender 90, 135
  pipes check 30
  tank 89, 90
Oil – transmission:-
  change 36, 92
  level check 31, 92
Oil pump:-
  bleeding 92
  cable adjustment 30
  removal and refitting 90, 91
  stroke adjustment 91, 92

P

Parking lamp 136, 137
Parts – ordering 6
Piston and rings:-
  examination and renovation 52
  refitting 66 – 68
  removal 44, 45
Primary drive gears:-
  examination and renovation 53
  refitting 60 – 62
  removal 46, 47
Power valve see **YPVS components**
Pulser coil 95
Pump – water 76, 77
Pump – oil 30, 90 – 92

R

Radiator 73 – 75
Rear brake – disc:-
  caliper 38, 120 – 121
  fluid renewal 36
  master cylinder 38, 123, 124
  pad check and renewal 32, 33
  pedal adjustment 34
Rear brake – drum:-
  adjustment 33, 34
  camshaft lubrication 37
  examination and renovation 125, 126
  pedal adjustment 34
Rear suspension:-
  specifications 98
  suspension unit 109, 110
  swinging arm and suspension linkage 35, 105 – 109
Rear wheel:-
  bearings 34, 35, 37, 119
  refitting 117, 118
  removal 116, 117
  sprocket and cush drive 119, 120
Reed valve:-
  examination 87, 88
  refitting 59, 60, 87, 88
  removal 49, 87, 88
Regulator/rectifier unit 133
Routine maintenance 23 – 38

S

Safety precautions 7
Seat and sidepanels 112
Side stand 35, 110, 134
Side stand switch 96
Source coil – ignition 94, 95
Spare parts – ordering 6

Spark plug:-
    check and adjustment 28, 97
    colour condition guide 29
    renewal 37
Specifications:-
    brakes 113
    clutch 40
    cooling system 71
    electrical system 131
    engine 39 – 41
    fuel system 78, 79
    gearbox 40, 41
    ignition system 93
    lubrication system 23, 78
    maintenance 23, 24
    suspension 98
    tyres 114
    wheels 113
Speedometer:-
    drive 37, 112
    drive cable 35, 112
    head 111
Sprocket:-
    gearbox 43
    rear wheel 119, 120
Steering:-
    check 35
    fault diagnosis 19
    head:
        bearings 37, 104, 105
        dismantling 103, 104
        reassembly 105
Stop lamp switch 33, 133
Stop/tail lamp 137, 138
Suppressor cap 97
Suspension:-
    check 35
    fault diagnosis 20
    front 36, 99 – 103
    rear 105 – 110
Swinging arm and suspension linkage:-
    check 35
    examination and renovation 106 – 108
    refitting 108, 109
    removal 105 – 107
Switches:-
    handlebar 133
    ignition and engine kill 95, 133
    neutral 49, 60, 110, 134
    side stand 35, 110, 134
    stop lamp 33, 133

T

Tachometer:-
    drive 48, 62, 112
    drive cable 35, 112
    head 110
Tank – fuel 79 – 81
Tap 81 – 82

Temperature gauge – coolant:-
    sender unit 77
    gauge circuit 135
Thermostat 75, 76
Throttle cable adjustment 30
Timing:-
    balancer shaft gears 60 – 62
    ignition 97
Tools 8 – 10
Torque settings 10, 24, 41, 71, 79, 98, 114, 131
Transmission:-
    fault diagnosis 18, 19
    oil change 36, 92
    oil level check 31, 92
Turn signal:-
    lamps 137, 138
    relay 134
Tyres:-
    check 25
    removal, repair and refitting:
        tubed 126, 127
        tubeless 126, 128, 129
    specifications 114
    valves 128, 130

V

Valve – power 46, 66
Valve – reed 49, 59, 60, 87, 88
Valve – tyre 128, 130
Voltage regulator 133

W

Water pump 76, 77
Weights – model 6
Wheel:-
    alignment 27
    balance 35, 130
    bearings 34, 35, 37, 118, 119
    check 34, 35
    removal and refitting:
        front 114 – 116
        rear 116 – 118
    specifications 113
    sprocket and cush drive – rear wheel 119, 120
Wiring diagrams 139 – 143
Working facilities 8, 9

Y

YEIS 88
YPVS components:-
    refitting 66
    removal 46